Modification of Selected South Carolina Bridge-Scour Envelope Curves

By Stephen T. Benedict and Andral W. Caldwell

Prepared in cooperation with the
South Carolina Department of Transportation

Scientific Investigations Report 2012–5029

U.S. Department of the Interior
U.S. Geological Survey

U.S. Department of the Interior
KEN SALAZAR, Secretary

U.S. Geological Survey
Marcia K. McNutt, Director

U.S. Geological Survey, Reston, Virginia: 2012

For more information on the USGS—the Federal source for science about the Earth, its natural and living resources, natural hazards, and the environment, visit *http://www.usgs.gov* or call 1-888-ASK-USGS

For an overview of USGS information products, including maps, imagery, and publications, visit http://www.usgs.gov/pubprod

To order this and other USGS information products, visit *http://store.usgs.gov*

Suggested citation:

Benedict, S.T., and Caldwell, A.W., 2012, Modification of selected South Carolina bridge-scour envelope curves: U.S. Geological Survey Scientific Investigations Report 2012-5029, 37 p., available online at *http://pubs.usgs.gov/sir/2012/5029/*.

Contents

Figures

Tables

Conversion Factors, Definitions, and Abbreviations

SI to Inch/Pound

Multiply	By	To obtain
Length		
millimeter (m)	0.3937	inch (in.)

Inch/Pound to SI

Multiply	By	To obtain
Length		
foot (ft)	0.3048	meter (m)
Area		
square mile (mi²)	2.590	square kilometer (km²)
Flow rate		
foot per second (ft/s)	0.03048	meter per second (m/s)
Hydraulic conductivity		
foot per foot (ft/ft)	0.3048	meter per meter (m/m)

Vertical coordinate information is referenced to North American Vertical Datum of 1929 or 1988 (NGVD 29 or NGVD 88, respectively).

Horizontal coordinate information is referenced to the North American Datum of 1927 or 1983 (NAD 27 or NAD 83, respectively).

Elevation, as used in this report, refers to distance above the vertical datum.

AEP	Annual Exceedance Probability
HEC-18	Hydraulic Engineering Circular No. 18
NBSD	National Bridge Scour Database
SCDOT	South Carolina Department of Transportation
USGS	U.S. Geological Survey
WSPRO	Water-Surface PROfile model

Modification of Selected South Carolina Bridge-Scour Envelope Curves

By Stephen T. Benedict and Andral W. Caldwell

Abstract

Historic scour was investigated at 231 bridges in the Piedmont and Coastal Plain physiographic provinces of South Carolina by the U.S. Geological Survey in cooperation with the South Carolina Department of Transportation. These investigations led to the development of field-derived envelope curves that provided supplementary tools to assess the potential for scour at bridges in South Carolina for selected scour components that included clear-water abutment, contraction, and pier scour, and live-bed pier and contraction scour. The envelope curves consist of a single curve with one explanatory variable encompassing all of the measured field data for the respective scour components. In the current investigation, the clear-water abutment-scour and live-bed contraction-scour envelope curves were modified to include a family of curves that utilized two explanatory variables, providing a means to further refine the assessment of scour potential for those specific scour components. The modified envelope curves and guidance for their application are presented in this report.

Introduction

The U.S. Geological Survey (USGS), in cooperation with the South Carolina Department of Transportation (SCDOT), investigated historic scour at 231 bridges in the Piedmont and Coastal Plain physiographic provinces of South Carolina (Benedict, 2003; Benedict and Caldwell, 2006; Benedict and Caldwell, 2009). [*Note:* The terms Piedmont and Coastal Plain are used in the remainder of the report to refer to the Piedmont and Coastal Plain physiographic

provinces, respectively. Historic scour refers to the maximum scour depth associated with a specific scour component, such as pier or abutment scour, that has occurred over the life of the bridge.] The general objectives of these studies were to (1) collect field measurements of historic abutment, contraction, and pier scour at sites that could be associated with major floods and (or) older bridges; (2) use the field data to assess the performance of the scour-prediction equations listed in the Federal Highway Administration Hydraulic Engineering Circular No. 18 (HEC-18; Richardson and Davis, 2001); and (3) develop supplementary tools derived from the field data to help assess scour potential in the Piedmont and Coastal Plain of South Carolina.

The results from previous investigations showed that HEC-18 scour-prediction equations generally overpredicted scour depths and were at times excessive. In some cases, however, substantial underprediction occurred, indicating that the equations could not be relied upon to consistently give conservative and reasonable estimates of scour. Although the HEC-18 equations provide a valuable resource for assessing scour, the results from the analysis highlighted the need for engineering judgment to determine if predicted scour is reasonable. To assist engineers in developing and applying such judgment, the data collected from the South Carolina field investigations were organized into regional bridge-scour envelope curves that displayed the range and trend for the upper limit of observed scour for each scour component, including clear-water abutment, clear-water contraction, and clear-water pier scour, as well as live-bed contraction and live-bed pier scour. [*Note:* Clear-water bridge scour in South Carolina typically occurs on the floodplain where flood-flow velocities are low and cannot transport bed sediments into the scour hole. Under clear-water scour conditions scour holes do not refill with sediments. In contrast, live-bed bridge scour in South Carolina typically occurs in the main channel where flood-flow velocities are sufficiently high to move bed sediments into the scour hole. Under live-bed scour conditions, scour holes are partially or totally refilled with sediments. Refer to Benedict (2003) and Benedict and Caldwell (2006, 2009) for additional information.] The envelope curves, for the respective scour components, consist of a single curve with one explanatory variable. Although the regional envelope curves have limitations (Benedict, 2003; Benedict and Caldwell, 2006, 2009), they can be used as a supplementary tool to evaluate predicted scour as well as the potential for scour at bridge sites in South Carolina.

The USGS and the SCDOT recently (2010) began a cooperative effort to test the application of the South Carolina bridge-scour envelope curves at selected bridges in South Carolina with unknown foundations. During the initial phase of that investigation it became apparent that the clear-water abutment-scour and live-bed contraction-scour envelope curves (Benedict, 2003; Benedict and Caldwell, 2009) could be modified to include a family of curves (or secondary envelope curves) that would provide a refined assessment of the upper limit of observed historic scour depths in South Carolina, in particular for sites associated with smaller drainage areas. [*Note:* The phrase "secondary envelope curve" is used in this report as a general term to refer to the individual envelope curves that make up the family of curves.] A review of the patterns in laboratory and field data for abutment and contraction scour led to the formulation of conceptual models for developing the modified envelope curves. Application of the conceptual models to the South Carolina field data and limited data from other locations in the United States produced a family of curves that can be used to assess the potential for clear-water abutment and live-bed contraction scour at bridges in the Piedmont and Coastal Plain of South Carolina. There are limitations associated with the modified envelope curves, and judgment must be used in their application. The purpose of this report is to describe the field data, conceptual models, and methods used in the development of the modified envelope curves, as well as their application and limitations.

Previous Investigations

The USGS, in cooperation with the SCDOT, has conducted five previous investigations of bridge scour in South Carolina. The first investigation was the level-1 bridge-scour project (1990–92), which included limited field-data collection at 3,506 bridges. These data were used to compute observed- and potential-scour indexes to identify bridges that may be susceptible to scour and require a more detailed analysis (Hurley, 1996). The second investigation was the level-2 bridge-scour project (1992–95), which included level-2 bridge-scour evaluations at 293 bridges using methods presented in HEC-18 (Richardson and Davis, 2001). The level-1 and level-2 bridge-scour studies gave a qualitative

overview of bridge scour in South Carolina, and indicated apparent discrepancies between scour depths in the field and scour depths predicted with the HEC-18 (Richardson and Davis, 2001) equations. These findings led to a series of three field investigations of historic bridge scour in South Carolina (Benedict, 2003; Benedict and Caldwell, 2006, 2009).

The field investigations of historic bridge scour collected data at 231 bridges in South Carolina that included 208 measurements of clear-water abutment scour, 189 measurements of clear-water pier scour, 139 measurements of clear-water contraction scour, 151 measurements of live-bed pier scour, and 89 measurements of live-bed contraction scour. The historic data represent the maximum scour depths at the time of the measurement that have occurred at the bridge since construction. The hydraulic conditions that produced the measured scour were approximated with a one-dimensional step-backwater model. The approximated hydraulic characteristics are less than ideal and introduce uncertainty in the data analysis, but the large number of data provides a means for assessing general field trends of scour in South Carolina. These data were used to evaluate the performance of the HEC-18 (Richardson and Davis, 2001) scour-prediction equations and to develop the previously noted bridge-scour envelope curves. The clear-water abutment-scour and live-bed contraction-scour envelope curves, as well as their associated data, were the primary focus for this investigation; a more detailed description of those data are presented later in the report.

Description of Study Area

South Carolina has an area of about 31,100 square miles (mi^2) and is divided into three physiographic provinces – the Blue Ridge, Piedmont, and Coastal Plain. The Coastal Plain province is divided into upper and lower regions (fig. 1). The study area for this investigation includes most of South Carolina but generally excludes the Blue Ridge and the tidally influenced area of the lower Coastal Plain.

The Piedmont covers about 35 percent of South Carolina and lies between the Blue Ridge and Coastal Plain (fig. 1). Land-surface elevations range from about 400 feet (ft) near the Fall Line (Coastal Plain boundary) to about 1,000 ft at the Blue Ridge boundary. The general topography includes rolling hills, elongated ridges, and moderately deep to shallow valleys. The drainage patterns are well developed with well-defined channels and densely vegetated floodplains. Streambed slopes in the Piedmont range from approximately 0.00015 to 0.0100 foot per foot (ft/ft) (Guimaraes and Bohman, 1992). The geology of the Piedmont generally consists of fractured crystalline rock overlain by moderately to poorly permeable silty-clay loams. Alluvial deposits along the valley floors generally consist of clay, silt, and sand, and form varying degrees of cohesive soils (Guimaraes and Bohman, 1992). The stream-channel sediments typically consist of sandy materials overlying decomposed rock or bedrock.

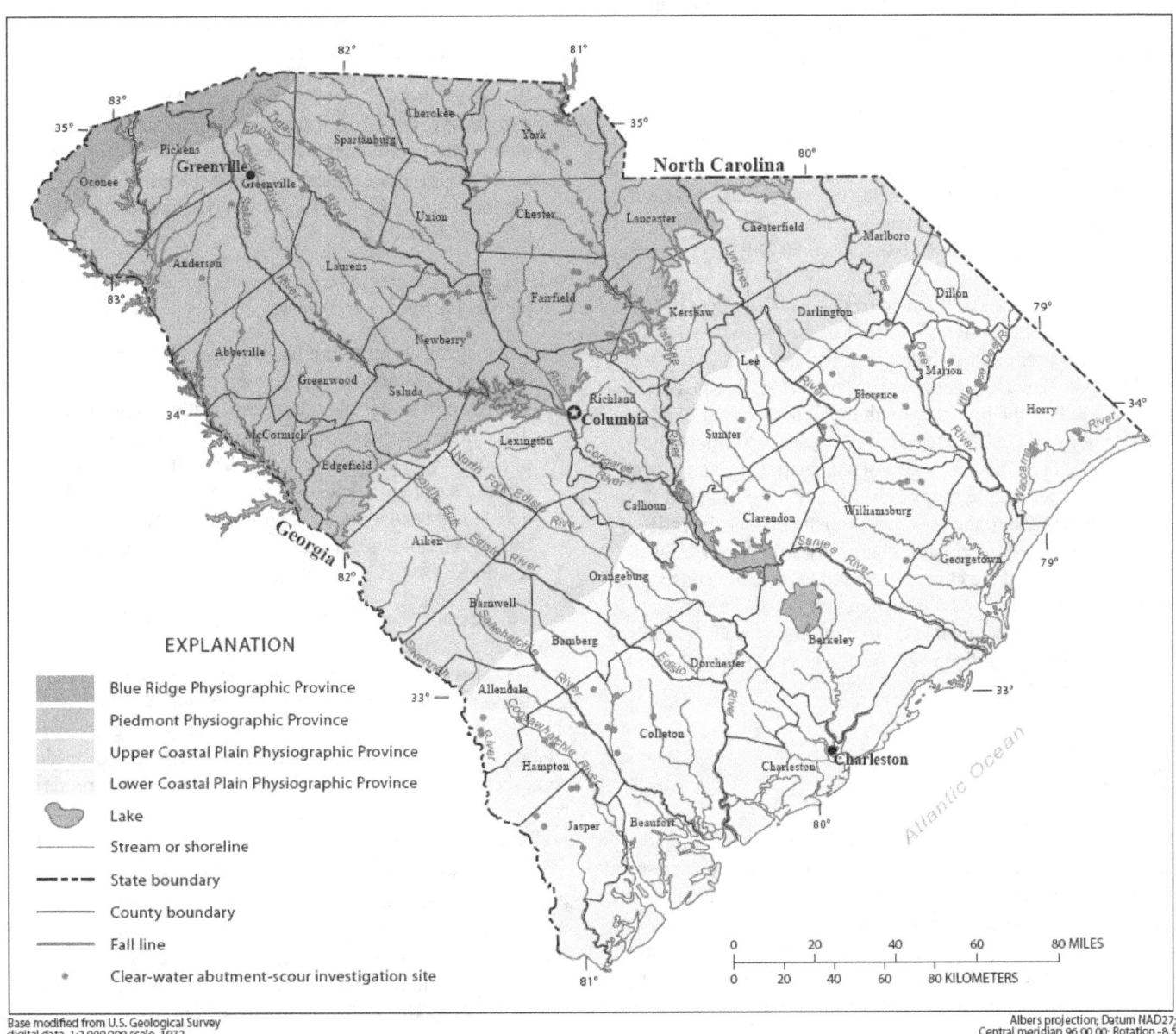

Figure 1. Location of selected clear-water abutment-scour investigation sites in South Carolina.

The upper Coastal Plain lies between the Piedmont and lower Coastal Plain, and covers about 20 percent of the State (fig. 1). The general topography in the upper Coastal Plain consists of rounded hills with gradual slopes, and land-surface elevations that range from about 200 ft to greater than 700 ft. The geology consists primarily of sedimentary rocks made up of layers of sand, silt, clay, and gravel underlain by igneous rocks (Zalants, 1990). A shallow surface layer of permeable sandy soils is common. Low-flow stream channels bounded by densely vegetated floodplains characterize upper Coastal Plain streams, and the channel sediments typically consist of sandy materials overlying rock. Streambed slopes are moderate, ranging from approximately 0.0005 to 0.0040 ft/ft (Guimaraes and Bohman, 1992).

The lower Coastal Plain covers about 43 percent of the State (fig. 1). The topographic relief in the lower Coastal Plain is less pronounced than that of the upper Coastal Plain, and land-surface elevations range from 0 ft at the coast to nearly 200 ft at the boundary with the upper Coastal Plain. The geology of the lower Coastal Plain consists of loosely consolidated sedimentary rocks of sand, silt, clay, and gravel overlain by permeable sandy soils (Zalants, 1991). As in the upper Coastal Plain, the low-flow stream channels bounded by densely vegetated floodplains characterize the lower Coastal Plain streams, and the channel sediments typically consist of sandy materials overlying sedimentary rock. Streambed slopes range from about 0.0001 to 0.0040 ft/ft, and streamflow patterns are tidally influenced near the coast (Guimaraes and Bohman, 1992).

Approach

Laboratory investigations of bridge scour have frequently used envelope curves to graphically display the trends of scour data and to develop semi-empirical relations for evaluating the potential for scour (Breusers and others, 1977; Breusers and Raudkivi, 1991; Dongol, 1993; Melville and Coleman, 2000; Ettema and others, 2011). With the current use of computers to model complex physical phenomena, the use of envelope curves for evaluating bridge scour seems too simplistic and somewhat archaic. However, the use of simple envelope curves, in large measure, stems from the limited understanding of the complex mechanisms that produce scour. The following quotations from selected researchers highlight this fact. In an extensive summary of bridge-scour research, Melville and Coleman (2000) report:

> The theoretical basis for the structural design of bridges is well established. In contrast, the mechanisms of flow and erosion in mobile-boundary channels have not been well defined and it is not possible to estimate with confidence the river boundary changes that may occur at a bridge subject to a given flood. This is not only due to the extreme complexity of the problem, but also to the fact that

river characteristics, bridge constriction geometry, and soil and water interaction are different for each bridge as well as for each flood.

In the findings of an extensive literature review of pier scour, Ettema and others (2011) report:

> Pier scour processes are intricate and challenging to formulate (even empirically or approximately), let alone fully comprehend. This statement holds for scour at all types of piers, especially those whose geometry consists of several components (column, pile cap, piles).

With respect to abutment scour, a more complex phenomenon than pier scour, Ettema and others (2005) report:

> Scant situations of hydraulic engineering are more complex than those associated with scour in the vicinity of a bridge abutment, especially one located in a compound channel. Accordingly, few situations of scour depth estimation are as difficult. Therefore, it is not surprising that considerable uncertainty and debate has been associated with scour depth estimation for abutments, and that the existing estimation relationships are not well accepted.

The limited understanding of the "extreme complexity" associated with bridge scour has necessitated the use of envelope curves for understanding scour trends in laboratory investigations and is a practice that likely will be associated with this discipline for years to come. Although envelope curves of laboratory data cannot provide a precise estimate of bridge scour, they are useful tools for defining upper-bound trends of scour, evaluating the influence of selected explanatory variables, and for developing semi-empirical scour-prediction equations.

Laboratory investigations have played a critical role in advancing the state-of-the-knowledge of bridge scour and will continue to do so in the future. However, there are continued concerns regarding small-scale laboratory investigations of bridge scour, including oversimplification of site conditions within the laboratory and scaling issues, both of which may lead to unreasonable estimates of scour when scaled to the field (Ettema and others, 1998; Ettema and others, 2004). One approach to minimizing these problems is to use field data, rather than laboratory data, to develop bridge-scour envelope curves. The use of field envelope curves can remove scaling problems associated with small-scale laboratory investigations and provide the practitioner with a better understanding of scour trends within the field setting. This was the approach used in the previous investigations to develop bridge-scour envelope curves for South Carolina (Benedict, 2003; Benedict and Caldwell, 2006, 2009). Numerous field measurements for each scour component were collected at bridges in the Piedmont and Coastal Plain of South Carolina, and dominant explanatory variables were used to develop envelope curves to define the upper bound of scour. In the previous investigations, the envelope curves consisted of a single curve with one explanatory variable encompassing all of the measured field data for the respective scour components. In the current investigation (2011), the clear-water abutment-scour (Benedict, 2003) and live-bed contraction-scour (Benedict and Caldwell, 2009) envelope curves were modified to include a family of curves with two explanatory variables, providing a means to further refine the assessment of scour potential for those specific scour components within the limits of the utilized data.

The Modified South Carolina Clear-Water Abutment-Scour Envelope Curves

The primary explanatory variables for abutment-scour depth as identified in laboratory investigations include flow duration, flow velocity, flow depth, sediment size, sediment gradation, embankment length, abutment shape, embankment skew, and channel geometry (Melville, 1992; Dongol, 1993; Melville and Coleman, 2000). [*Note:* Many laboratory investigations define the road embankment that blocks approaching flow as the abutment length. In this report, the term "embankment length" will be used to represent this variable.] Benedict (2003) reviewed the influence of each of these variables on the South Carolina clear-water abutment-scour field data (table 1)

and concluded that "many of these variables likely have minimal influence on abutment-scour depths for the prevailing field conditions in South Carolina." Benedict (2003) further concluded that the geometric variables of embankment length blocking flow (also called embankment length or abutment length) and the geometric-contraction ratio (m) were the strongest explanatory variables for the field data and noted that this conclusion was substantiated by laboratory investigations (Das, 1973; Melville, 1992; Dongol, 1993; Melville and Coleman, 2000; $m = 1 - b/B$, where b is the flow top width in the bridge opening and B is the flow top width at the upstream approach cross section). In particular, Dongol (1993) stated that, "Abutment length is one of the most important parameters influencing the process of local abutment scour." Similarly, Das (1973) concluded that the geometric-contraction ratio is an important parameter influencing abutment scour. Based on these findings, Benedict (2003) developed clear-water abutment-scour envelope curves that encompassed the upper bound of historic abutment scour in South Carolina using embankment length and the geometric-contraction ratio as explanatory variables (figs. 2–5). Because of the distinct regional characteristics of the Piedmont and Coastal Plain streams (see table 1) separate envelope curves were developed for each region. Although the abutment-scour envelope curves have limitations (Benedict, 2003) they can be used as a supplementary tool to evaluate predicted abutment scour as well as the potential for abutment scour at bridge sites in South Carolina.

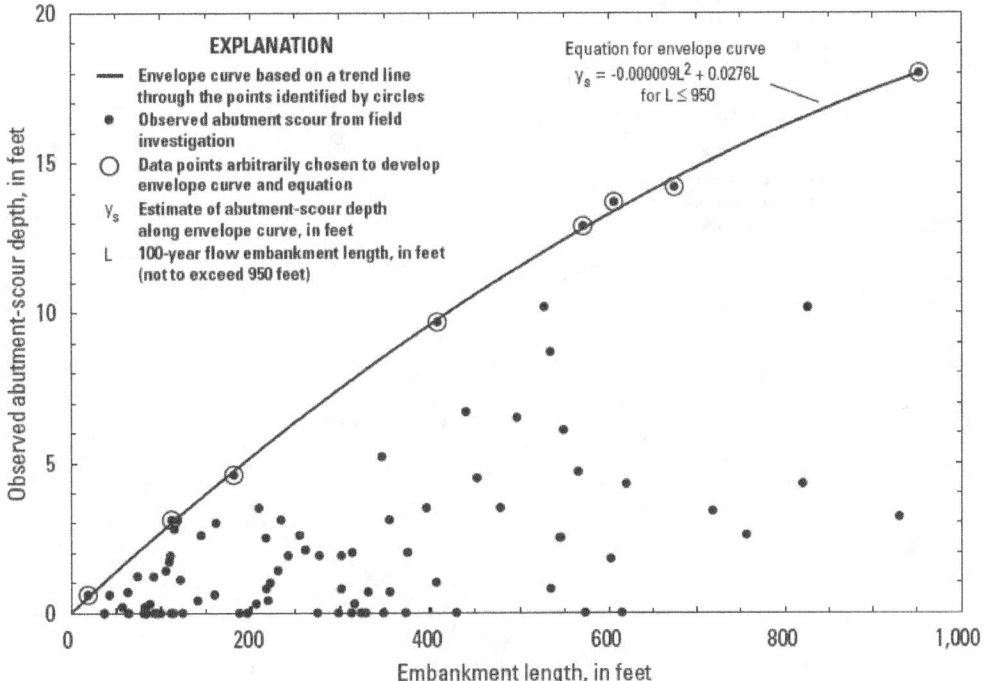

Figure 2. Relation of measured clear-water abutment-scour depth to embankment length in the Piedmont of South Carolina (from Benedict, 2003).

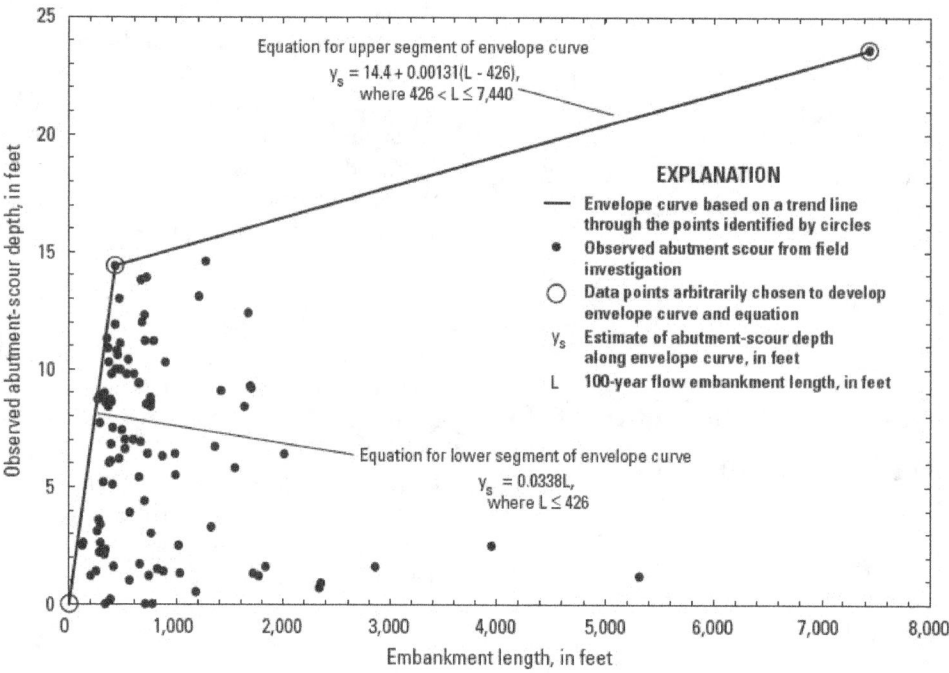

Figure 3. Relation of measured clear-water abutment-scour depth to embankment length in the Coastal Plain of South Carolina (from Benedict, 2003)

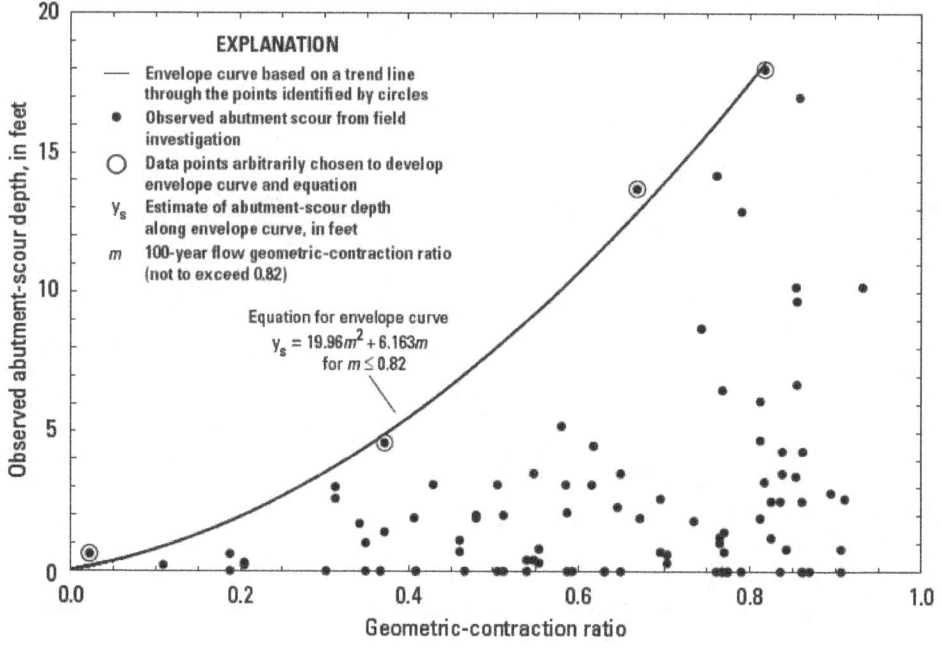

Figure 4. Relation of measured clear-water abutment-scour depth to the geometric-contraction ratio in the Piedmont of South Carolina (from Benedict, 2003).

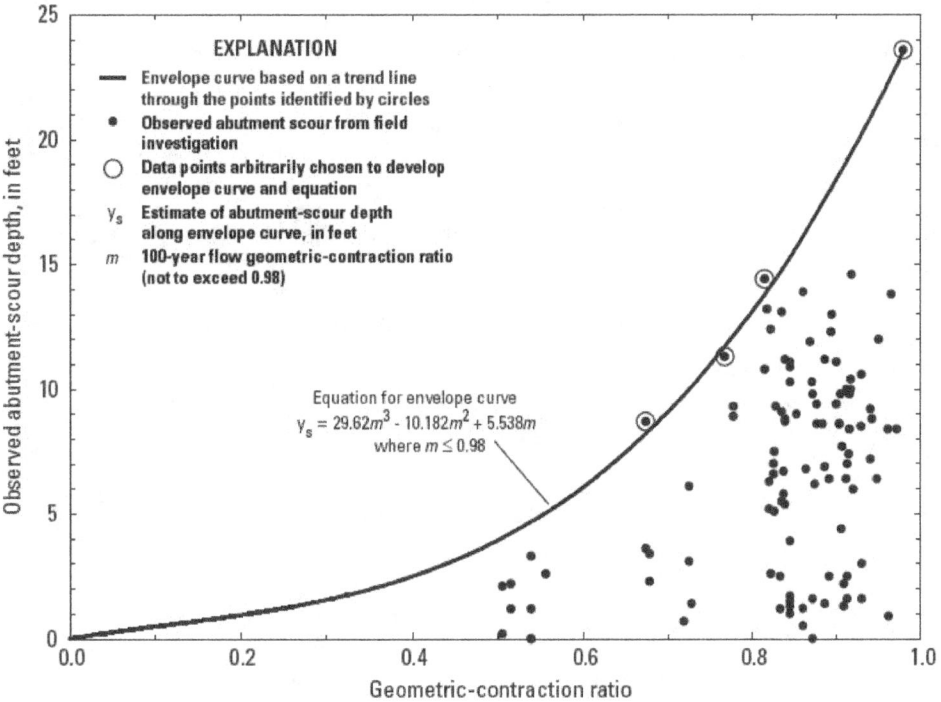

Figure 5. Relation of measured clear-water abutment-scour depth to the geometric-contraction ratio in the Coastal Plain of South Carolina (from Benedict, 2003).

An objective of the current (2011) investigation is to modify the existing clear-water abutment-scour envelope curves (Benedict, 2003) that utilize a single explanatory variable and to include a family of curves that utilize two explanatory variables, which will provide a means to refine the assessment of the upper bound of historic abutment scour in South Carolina. The following report sections document the development of the modified clear-water abutment-scour envelope curves with particular focus on (1) the field data used in the analysis, (2) the conceptual model, (3) the procedures used to develop the modified envelope curves, and (4) the guidance for applying the modified envelope curves.

Selected Field Data Used in the Analysis

The collection of field measurements pertaining to abutment scour at selected bridges in the United States was initiated in the late 1990s by the USGS in cooperation with other highway agencies. These data were collected in part to form a database to evaluate the performance of scour-prediction equations and to understand abutment-scour trends in the field setting. Currently available USGS abutment-scour data include 208 measurements in South Carolina (Benedict, 2003), 29 measurements in the USGS National Bridge Scour Database (NBSD; *http://water.usgs.gov/osw/techniques/bs/BSDMS/index.html*, accessed July 22, 2011; U.S. Geological Survey,

2001; Wagner and others, 2006), and 100 measurements in Maine (Lombard and Hodgkins, 2008). The South Carolina database was the primary source for developing the original (Benedict, 2003) clear-water abutment-scour envelope curves and also served as the primary data source for developing the modified envelope curves in this investigation. The Maine data are associated with stream characteristics distinctly different from the South Carolina data and, therefore, were not used in this investigation. The NBSD data, however, do have similar stream characteristics to those associated with the South Carolina data, and these were used for limited confirmation of the patterns of the South Carolina data. A brief description of these data and their limitations follows.

South Carolina Clear-Water Abutment-Scour Data

The South Carolina data includes 208 measurements of historic clear-water abutment scour collected at 138 bridges in the Piedmont and Coastal Plain of South Carolina (Benedict, 2003) with scour depths ranging from 0 to 23.6 ft (fig. 1; table 1). These scour measurements are assumed to represent the maximum clear-water abutment-scour depth that has occurred on the bridge overbanks since construction. All measurements were made on the floodplain at the bridge (also called the bridge overbank) and the reference surface used to determine the scour depth was the average undisturbed

Table 1. Range of selected site characteristics for field measurements of abutment scour.

[mi², square mile; ft/ft, foot per foot; ft/s, foot per second; ft, foot; mm, millimeter; <, less than; —, missing data]

Range value	Drainage area (mi²)	Channel slope (ft/ft)	Average blocked approach velocity (ft/s)	Average blocked approach depth (ft)	Embankment length blocking flow (ft)	Geometric-contraction ratio	Median grain size (mm)	Measured abutment-scour depth (ft)
South Carolina Piedmont (Benedict, 2003) (100 measurements)								
Minimum	11	0.00015	0.1	1.0	18	0.02	< 0.062	0.0
Median	75	0.0012	0.9	5.4	276	0.69	0.073	1.3
Maximum	1,620[a]	0.0029	3.2	14.6	953[b]	0.96	0.99	18.0
South Carolina Coastal Plain (Benedict, 2003) (108 measurements)								
Minimum	6	0.00007	0.1	1.5	127	0.51	< 0.062	0.0
Median	120	0.0005	0.5	4.7	631	0.86	0.18	7.0
Maximum	8,830[c]	0.0024	1.6	17.4	7,440[d]	0.98	0.78	23.6
National Bridge Scour Database (U.S. Geological Survey, 2001) (29 measurements)								
Minimum	836[e]	0.00006[e]	—	—	8[f]	0.41[f]	0.001[g]	0.0
Median	970[e]	0.0005[e]	—	—	527[f]	0.91[f]	0.15[g]	3.0
Maximum	16,000[e]	0.0046[e]	—	—	1,775[f]	0.93[f]	35.0[g]	18.0

[a]About 97 percent of the study sites in the Piedmont have drainage areas less than 400 mi².

[b]Three observations had embankment lengths exceeding 950 ft and were significantly outside the range for the majority of the Piedmont data. These sites were excluded from development of the embankment-length envelope curve.

[c]About 80 percent of the study sites in the Coastal Plain have drainage areas less than 426 mi².

[d]Only seven observations have embankment lengths greater than 2,000 ft.

[e]Data are missing for 1 measurement.

[f]Data are missing for 4 measurements.

[g]Data are missing for 12 measurements.

floodplain elevation in the vicinity of the observed scour. Because of clear-water scour conditions, infill sediments within the scour holes were, in general, negligible. The dominant abutment geometry was the spill-through abutment, which was observed at 135 bridges. The remaining 3 bridges had vertical wingwall abutments. A grab sample of sediment was obtained in the upstream floodplain at each site and was analyzed to estimate the median grain size. Because sediment characteristics in the field setting can vary substantially in the vertical and horizontal direction, the grab sample taken at a point may not fully represent the sediment characteristics at a site.

The South Carolina clear-water abutment-scour depths were measured during low flows and the flow conditions that produced the scour are not known. To estimate the hydraulic characteristics that may have produced the observed abutment scour, numerical models were developed for each site using the one-dimensional step-backwater model, Water-Surface PROfile (WSPRO; Shearman, 1990). A review of historic floods in South Carolina and a risk analysis associated with the bridge age, indicated that about 90 percent of the bridges in the study likely had experienced flows equaling or exceeding 0.7 times the 1-percent annual exceedance probability (AEP) flow (100-year flow) (Benedict, 2003). Therefore, the 1-percent AEP flow was assumed to be representative of a common flow that may have occurred at all bridges, and this flow was used in the WSPRO model to estimate the flow characteristics that may have produced the measured scour. In addition, historic flow records from streamflow gages were available at or near 35 bridges, and the maximum historic flow at these bridges also was used in the WSPRO model. All hydraulic characteristics associated with the South Carolina clear-water abutment-scour data were derived from the WSPRO model and should be viewed as approximate rather than measured data. The hydraulic characteristics approximated with a one-dimensional model may introduce error into computations and analysis associated with the abutment-scour data, making such analysis less than ideal. However, the large number of field measurements (208) in the database, provides insight into regional trends for abutment scour.

The clear-water abutment-scour data collected in South Carolina were grouped into two databases based on regional location within the State. One database contained data collected in the Piedmont physiographic province and the other contained data collected in the Coastal Plain physiographic province. [*Note:* The Coastal Plain database did not include bridges that were tidally influenced during high flows.] This division of the data was justified because of the distinct regional characteristics associated with the streams of the Piedmont and Coastal Plain Provinces. The Piedmont generally has cohesive floodplain soils, moderate stream gradients, relatively narrow floodplains, and relatively short flood-flow durations. In contrast, the Coastal Plain generally has sandy floodplain soils, low gradient streams, relatively wide floodplains, and relatively long flood-flow durations. To provide some understanding of the differences between these regions, table 1 lists the median and range of selected site characteristics for the clear-water abutment-scour field data collected in the Piedmont and Coastal Plain. For additional details regarding the South Carolina clear-water abutment-scour data, refer to Benedict (2003).

National Bridge Scour Database

The NBSD contains 29 measurements of abutment scour taken at various bridge sites throughout the United States. A review found that 25 of the abutment-scour measurements had sufficient supporting data that could be used to compare with patterns observed in the original South Carolina bridge-scour envelope curves (Benedict, 2003; figs. 2–5). However, 9 of these measurements had embankment lengths outside the range of the South Carolina data used in the current investigation, leaving only 16 measurements applicable for verifying the modified South Carolina abutment-scour envelope curves. Most of the data in the NBSD were collected during high-flow events, and measurements of the flows that produced the scour usually were taken concurrently with the scour measurements. Although the number of data from the NBSD is much smaller than the South Carolina dataset, it offers a valuable resource for comparison with the patterns observed in the South Carolina data. Because field measurements of flow in the NBSD were typically obtained at the bridge instead of the approach, one-dimensional hydraulic models were used to estimate hydraulic characteristics at the upstream approach section. This reliance on modeled flow properties will introduce some error into the computations and analysis associated with the abutment-scour data, making the analysis less than ideal. The median and range of selected site characteristics for the NBSD abutment-scour field data are listed in table 1. Additional details associated with the data can be found at the NBSD web page (*http://water.usgs.gov/osw/techniques/ bs/BSDMS/index.html*, accessed July 22, 2011) as well as in Wagner and others (2006).

Conceptual Model for the Modified Clear-Water Abutment-Scour Envelope Curves

The conceptual model used for this study to modify the original South Carolina clear-water abutment-scour envelope curves, assumes that (1) the geometric variables of embankment length blocking flow and geometric-contraction ratio are strong explanatory variables for abutment scour and can be utilized to develop envelope curves for the upper bound of historic abutment scour as demonstrated by Benedict (2003; figs. 2–5), and (2) these two explanatory variables can be utilized in a common envelope curve to develop a family of curves. The family of curves would consist of abutment-scour data grouped into categories according to embankment length and plotted against the geometric-contraction ratio. Support for this approach can be substantiated by a brief review of selected laboratory and field data.

As noted previously, the primary variables that influence abutment-scour depth include flow duration, flow velocity, flow depth, sediment size, sediment gradation, embankment length, abutment shape, embankment skew, and channel geometry (Melville, 1992; Dongol, 1993; Melville and Coleman, 2000). Although the interaction of these variables in producing abutment scour can be complex, making it difficult to determine the influence of an individual variable, laboratory investigations of abutment scour can be constructed in such a way as to more readily isolate the influence of a selected variable. If a laboratory dataset consists of (1) equilibrium-scour depths at threshold conditions (approach flow velocity is equal to the sediment critical velocity); (2) similar abutment and channel geometry; and (3) relatively fine uniform sediment sizes, such that the ratio of the embankment length to the median sediment size is 50 or greater, then the influence of flow duration, flow velocity, sediment size, sediment gradation, abutment shape, embankment skew, and channel geometry are minimized, thereby making embankment length and flow depth the primary influencing variables. Using selected laboratory data from Melville (1992) and Dongol (1993) that largely meet the above criteria, the general influence of embankment length on abutment-scour depth can be observed (figs. 6 and 7). Figure 6 includes 91 of the 96 threshold abutment-scour data from Melville (1992) along with 10 measurements from Dongol (1993) associated with longer embankment lengths than those of the Melville (1992) data. The pattern observed in figure 6 indicates that the upper bound of abutment-scour depth increases with increasing embankment length. The data in figure 7 (a subset of the data from fig. 6), were grouped by abutment shape and flow depth so as to isolate the influence of embankment length on abutment-scour depth for each group. Based on the patterns observed in figures 6 and 7, laboratory researchers have noted that abutment-scour depth asymptotically increases with increasing embankment length and approaches a limit where embankment length has minimal influence on abutment-scour depth (Melville, 1992; Dongol, 1993; Melville and Coleman, 2000). This pattern also is observed in the upper bound of the

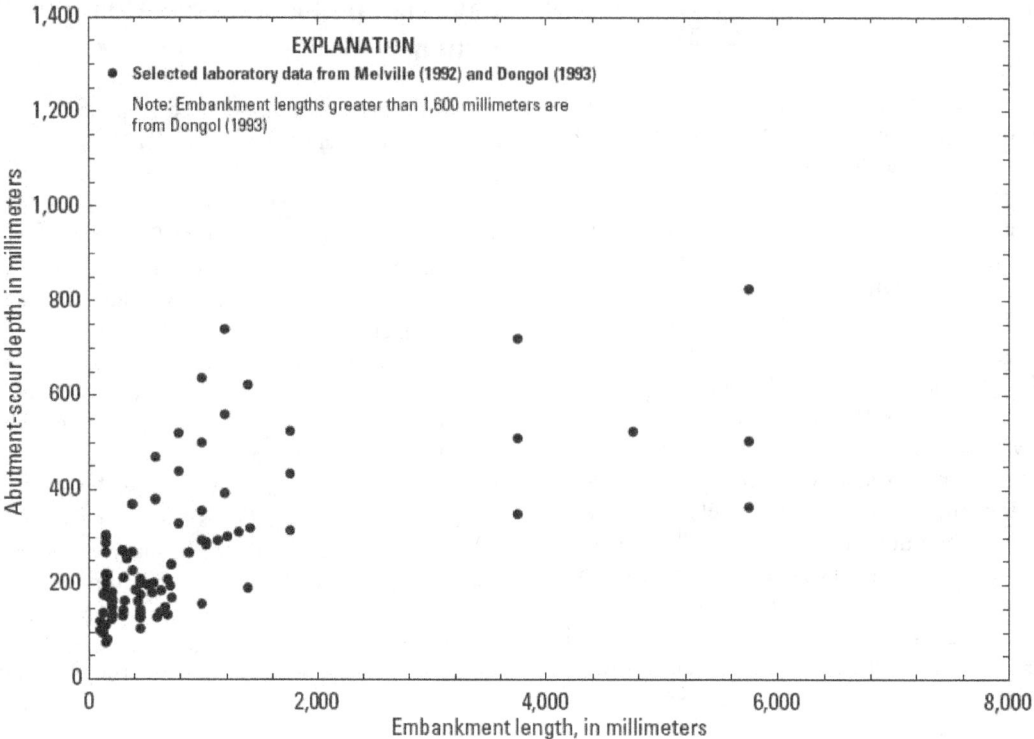

Figure 6.   Relation of abutment-scour depth to embankment length for selected laboratory data.

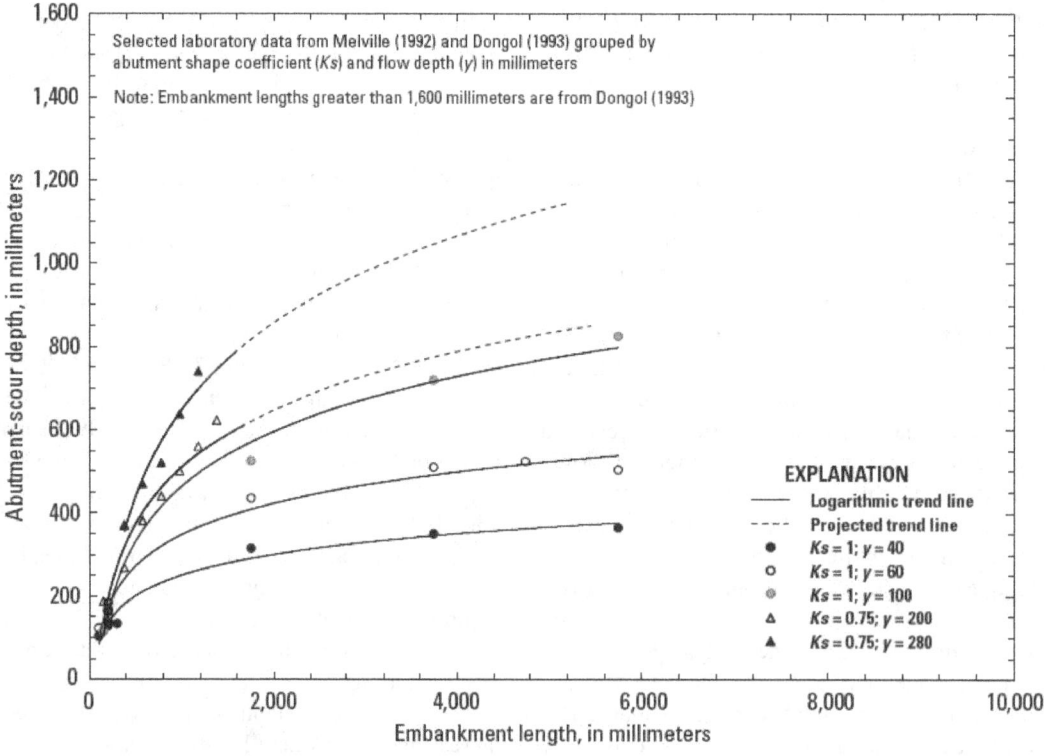

Figure 7.   Relation of abutment-scour depth to embankment length for selected laboratory data grouped by shape and flow depth.

South Carolina data (Benedict, 2003; figs. 2 and 3) as well as in the NBSD data (U.S. Geological Survey, 2001; fig. 8). Figure 8 includes the original South Carolina abutment-scour envelope curves (Benedict, 2003; figs. 2 and 3) and shows how the upper bound of the NBSD data closely conforms to those envelopes, indicating that the patterns observed in the South Carolina data are similar to those found in other areas of the country.

The selected Melville (1992) and Dongol (1993) laboratory data also can be used to display the general influence of the geometric-contraction ratio on abutment-scour depth (figs. 9 and 10). These data indicate that the upper bound of abutment-scour depth increases with increasing geometric-contraction ratio (fig. 9). The grouped data in figure 10 show the influence of the geometric-contraction ratio on abutment-scour depth and further confirms this trend. Das (1973) observed a similar pattern (fig. 11) and concluded that the geometric-contraction ratio was a strong explanatory variable for abutment-scour depth.

The clear-water abutment-scour field data from South Carolina (Benedict, 2003; figs. 4 and 5) and the NBSD (U.S. Geological Survey, 2001; fig. 12) display similar patterns to those shown in the laboratory data. In particular,

the upper bound of the field data displays the same pattern as the Das (1973) data, indicating that abutment-scour depth will increase with increasing geometric-contraction ratio. Figure 12 includes the original South Carolina abutment-scour envelope curves (Benedict, 2003; figs. 4 and 5) and shows that the NBSD data closely conform to those envelopes, indicating that the patterns observed in the South Carolina data are similar to those found in other areas of the country.

Previous laboratory studies of abutment scour have typically not investigated the combined effect of embankment length and the geometric-contraction ratio on abutment-scour depth (Ballio and others, 2009). To investigate the combined effect of these variables on abutment-scour depth, laboratory investigations must be constructed so that the embankment length and other influencing variables are held constant while varying the geometric-contraction ratio. To vary the geometric-contraction ratio in such a manner requires that the walls of the laboratory flume be adjusted while maintaining a constant embankment length. However, laboratory investigations of abutment scour typically maintain a constant flume width while varying the embankment length, thus limiting the laboratory data currently (2011) available for defining the relation between these two variables with respect to

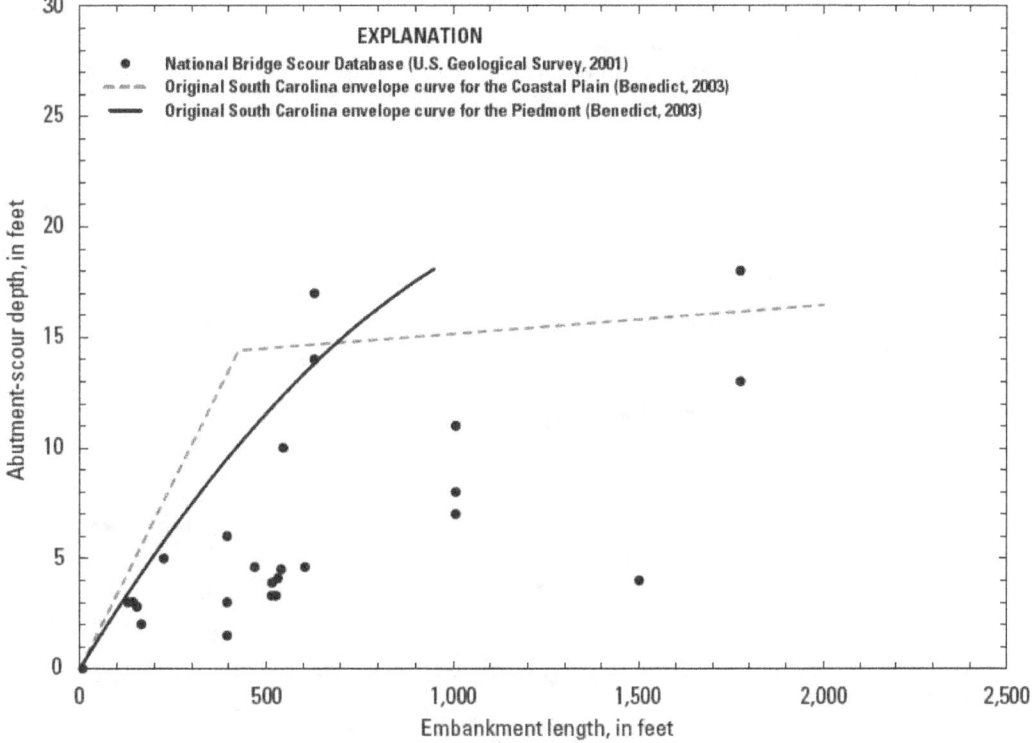

Figure 8. Relation of abutment-scour depth to embankment length for field data from the National Bridge Scour Database and the South Carolina envelope curves.

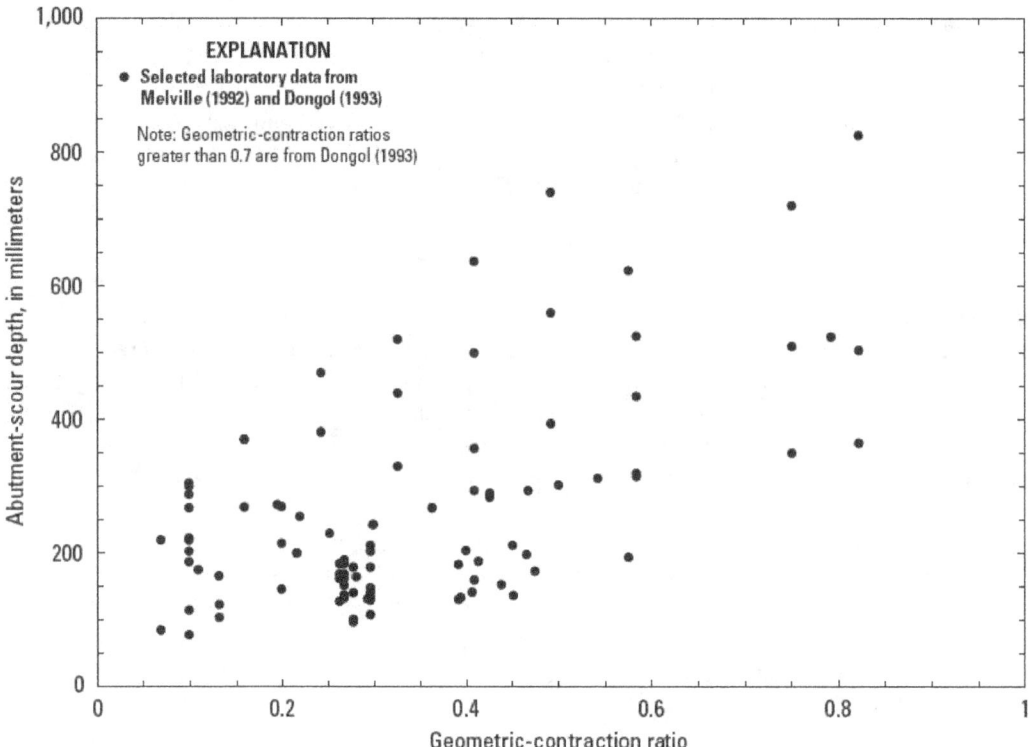

Figure 9. Relation of abutment-scour depth to the geometric-contraction ratio for selected laboratory data.

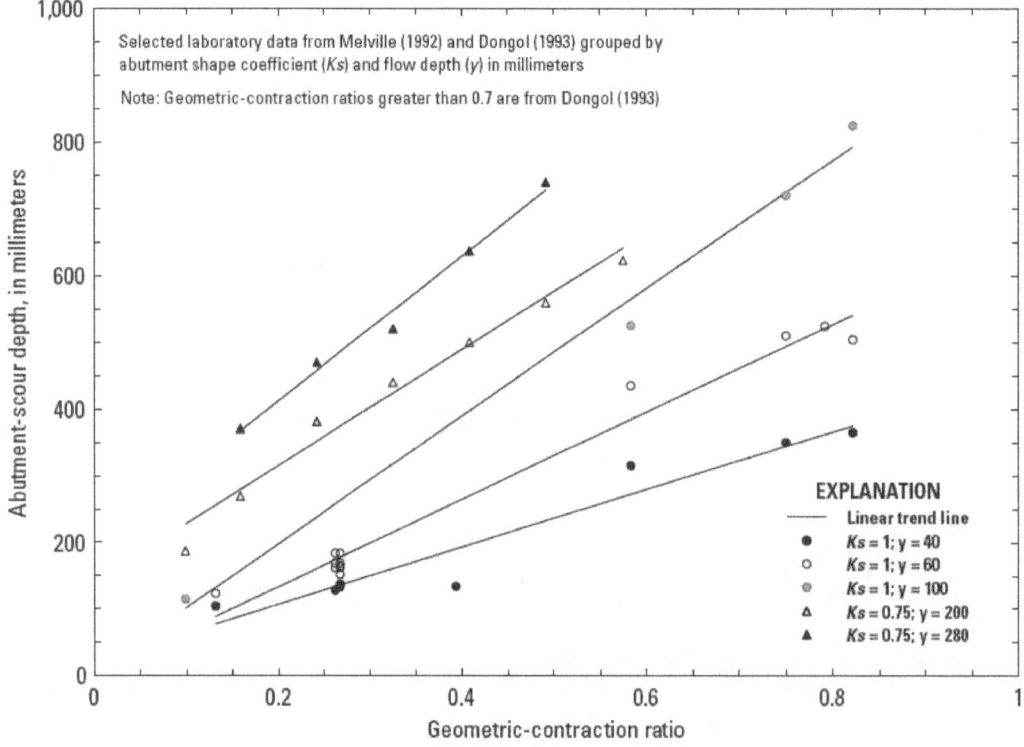

Figure 10. Relation of abutment-scour depth to the geometric-contraction ratio for selected laboratory data grouped by shape and flow depth.

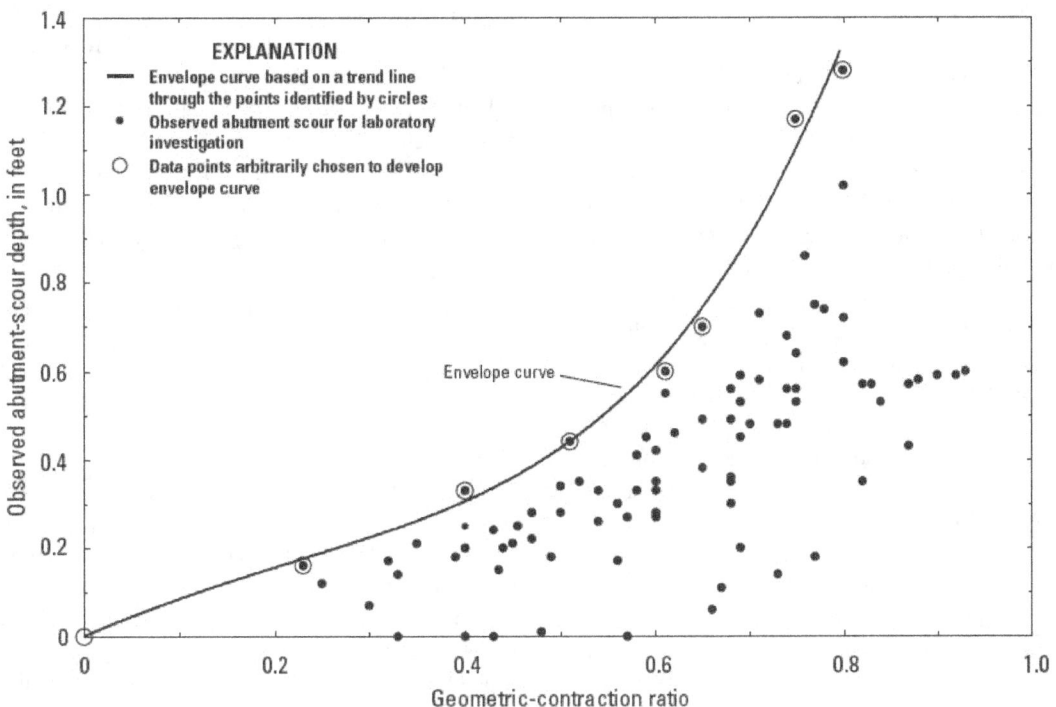

Figure 11.   Relation of abutment-scour depth to the geometric-contraction ratio for laboratory data. (Das, 1973; plot from Benedict, 2003).

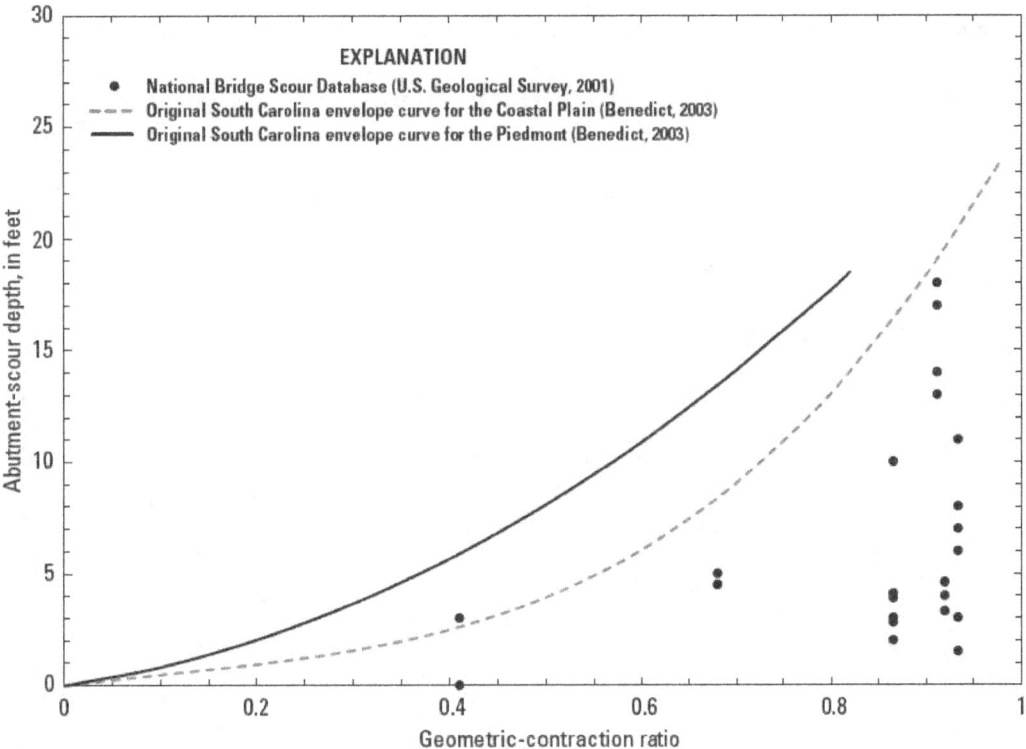

Figure 12.   Relation of abutment-scour depth to the geometric-contraction ratio for field data from the National Bridge Scour Database and the South Carolina envelope curves.

abutment-scour depth. To better understand this relation, Ballio and others (2009) conducted a limited laboratory investigation in which the geometric-contraction ratio was varied for a constant embankment length. These data were grouped by embankment length and flow depth to isolate the effect of the geometric-contraction ratio and are shown in figure 13.

Based on these laboratory data, Ballio and others (2009) noted that for geometric-contraction ratios of approximately 0.33 or less, the effect on local abutment-scour depth was negligible. Breusers and Raudkivi (1991) made a similar observation but recommend an upper limit of the geometric-contraction ratio to be 0.4. This general pattern is also observed in the Das (1973) laboratory data (fig. 11) where geometric-contraction ratios of approximately 0.4 or less do not influence the upper bound of abutment scour as substantially as larger values. Ballio and others (2009; fig. 13) also noted that local abutment-scour depth can increase substantially for geometric-contraction ratios greater than 0.33, and this observation is substantiated by the Das (1973; fig. 11) data. It is noteworthy that the upper bound of the South Carolina (Benedict, 2003; figs. 4 and 5) and the NBSD (U.S. Geological Survey, 2001; fig. 12) abutment-scour data have similar patterns to the laboratory data, providing increased confidence that the patterns displayed in the field data are reasonable. The patterns in figure 13 demonstrate that the relation of abutment-scour depth to the geometric-contraction ratio will form distinct curves for a constant embankment

length, indicating that it may be possible to develop a family of curves using field data that show the relation of abutment-scour depth to the geometric-contraction ratio for a series of selected embankment-length categories.

Based on the above-mentioned observations from the laboratory and field data, the following conceptual model was used in the development of the modified South Carolina clear-water abutment-scour envelope curves:

For a constant embankment length, abutment-scour depth will increase at an increasing rate as the geometric-contraction ratio increases.

The conceptual model is illustrated in figure 14 by a typical laboratory configuration in which a single embankment is placed at the wall of a rectangular flume, and represents half of a bridge opening that is symmetrical about the wall opposite of the abutment. In this illustration (fig. 14), the embankment length blocking flow (L) is held constant while being subjected to a series of increasing geometric-contraction ratios (scenarios 1–3) that produce increasing abutment-scour depths. The generalized trend for abutment-scour depth associated with the scenarios in figure 14 is shown in figure 15. If abutment-scour data can be grouped by categories of constant embankment length that have varying geometric-contraction ratios, then a family of curves similar to those

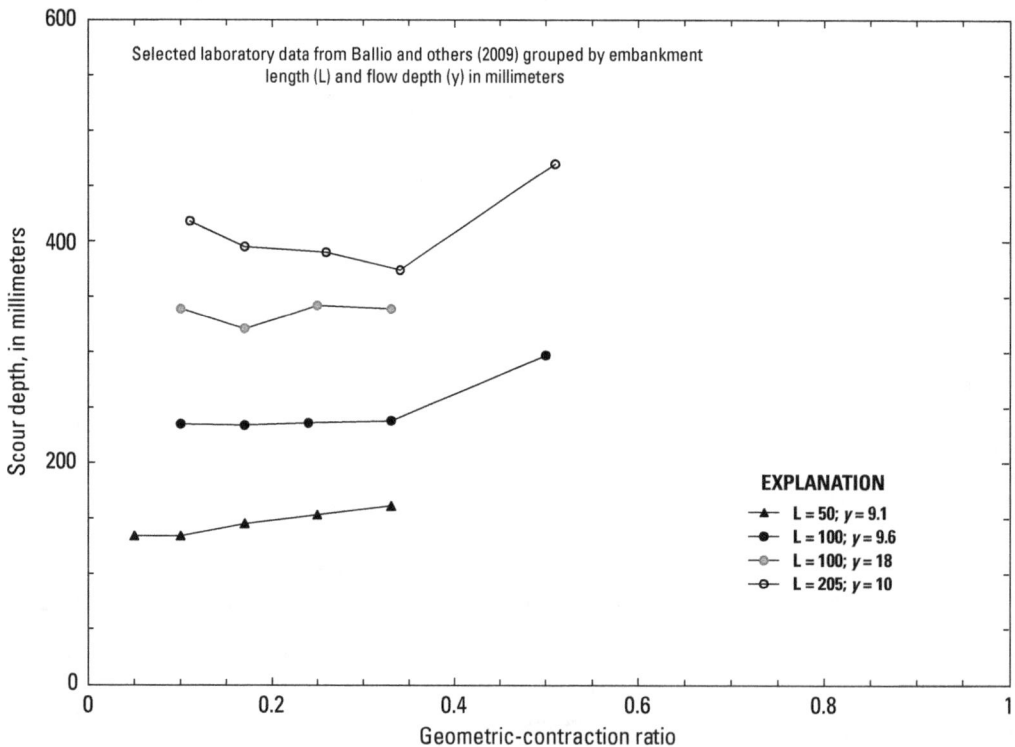

Figure 13. Relation of abutment-scour depth to the geometric-contraction ratio for selected laboratory data grouped by embankment length and flow depth.

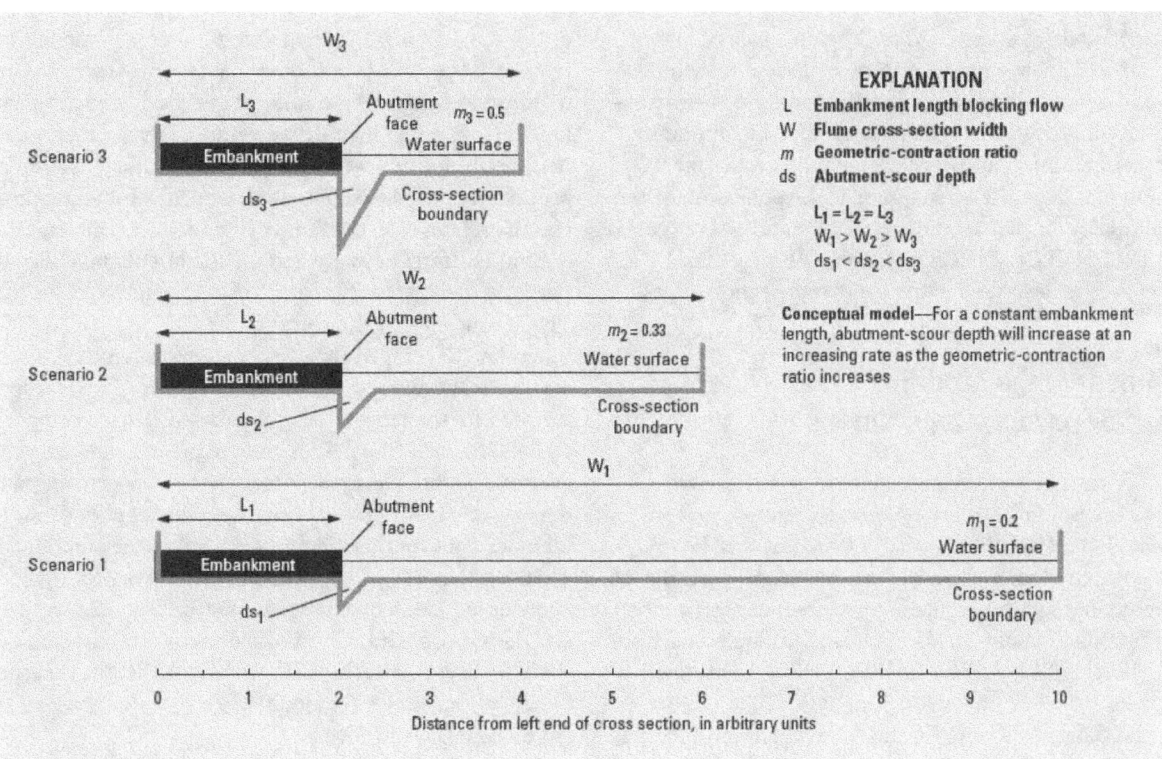

Figure 14.   Schematic of the conceptual model for the relation of abutment-scour depth to the geometric-contraction ratio for a constant embankment length.

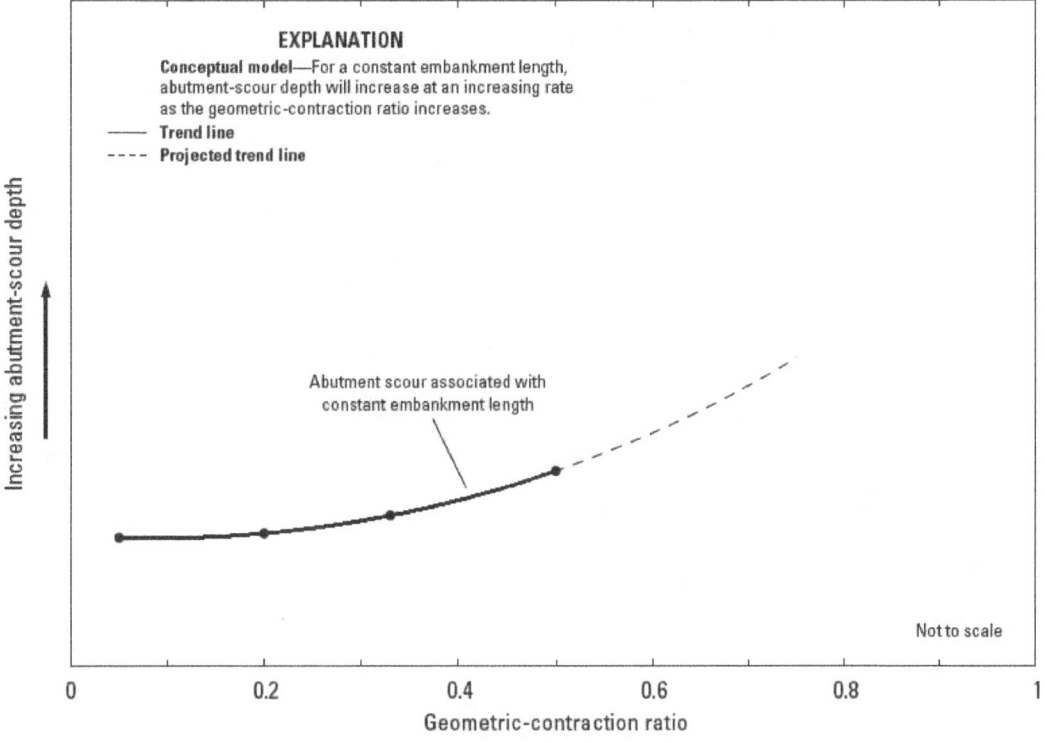

Figure 15.   Generalized trend for the relation of abutment-scour depth to the geometric-contraction ratio for a constant embankment length.

in figure 16 can be developed. [*Note:* The presence of an embankment will always produce some contraction such that the minimum geometric-contraction ratio for a given embankment length will always be greater than zero. The minimum geometric-contraction ratio will increase with increasing embankment length as shown in figure 16.] Application of this conceptual model for the development of a family of curves to modify the South Carolina clear-water abutment-scour envelope curves is described in the following section.

Development of the Modified Clear-Water Abutment-Scour Envelope Curves

To apply the conceptual model that was previously described to the South Carolina clear-water abutment-scour data and develop a family of curves, the ideal would be to have a series of field measurements associated with a constant embankment length with varying geometric-contraction ratios. But obtaining field data in such a manner is difficult and, from a practical view, unlikely. Therefore, the data were grouped into categories based on embankment length. Although the number of field data available for this investigation was large

(224), the data were limited and, for a given embankment-length category, data could be sparse. Therefore, judgment was required to construct a family of curves that encompassed the field data and in which the secondary envelope curves were reasonably located with respect to each other. The following process was used to define the categories for embankment length and develop the family of curves. Because of the regional differences between the Piedmont and Coastal Plain (table 1), data from these regions were analyzed separately. The secondary envelope curves for the Piedmont and Coastal Plain data developed for each category of embankment length are shown in figures 17 through 21. [*Note:* Each figure shows the data for the respective embankment-length category as well as the data for smaller embankment-length categories. For example, figure 19 represents the secondary envelope curves developed for the embankment-length category of 300 ft and includes data for that category as well as the smaller categories of 100 and 200 ft.] The full family of curves for the Piedmont and Coastal Plain are shown in figures 22 and 23, respectively, and demonstrate how each of the secondary envelope curves spatially relate to each other. Additional details regarding the development of the modified clear-water abutment-scour envelope curves follow.

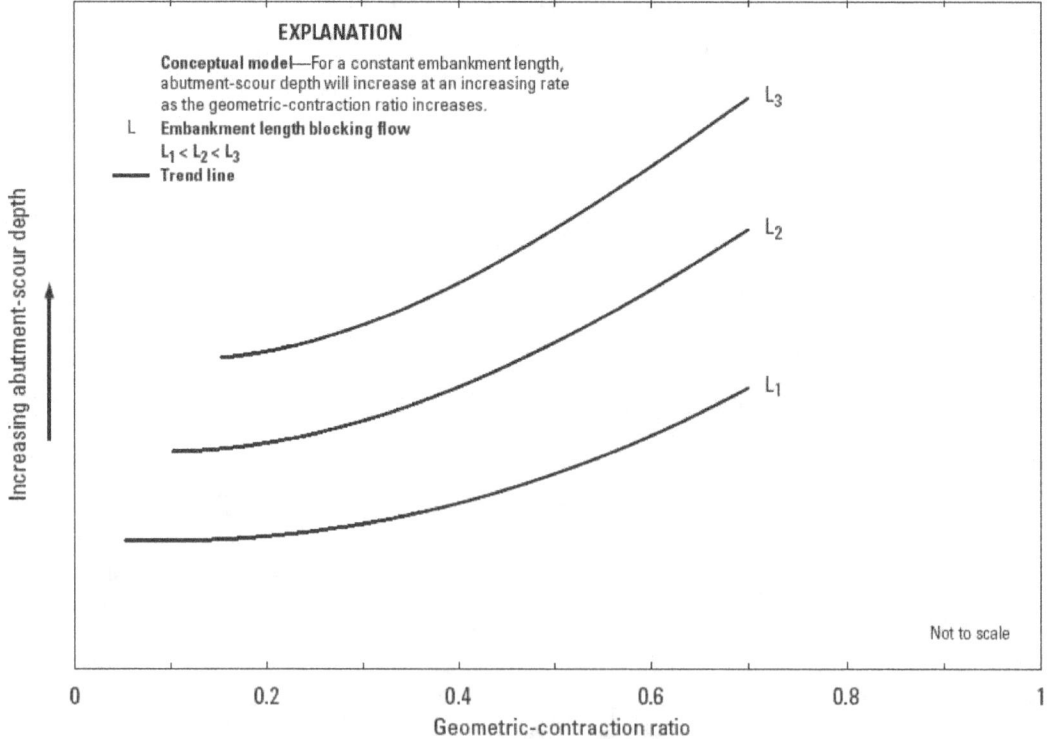

Figure 16.   Family of curves for the generalized trend of the relation of abutment-scour depth to the geometric-contraction ratio for a constant embankment length.

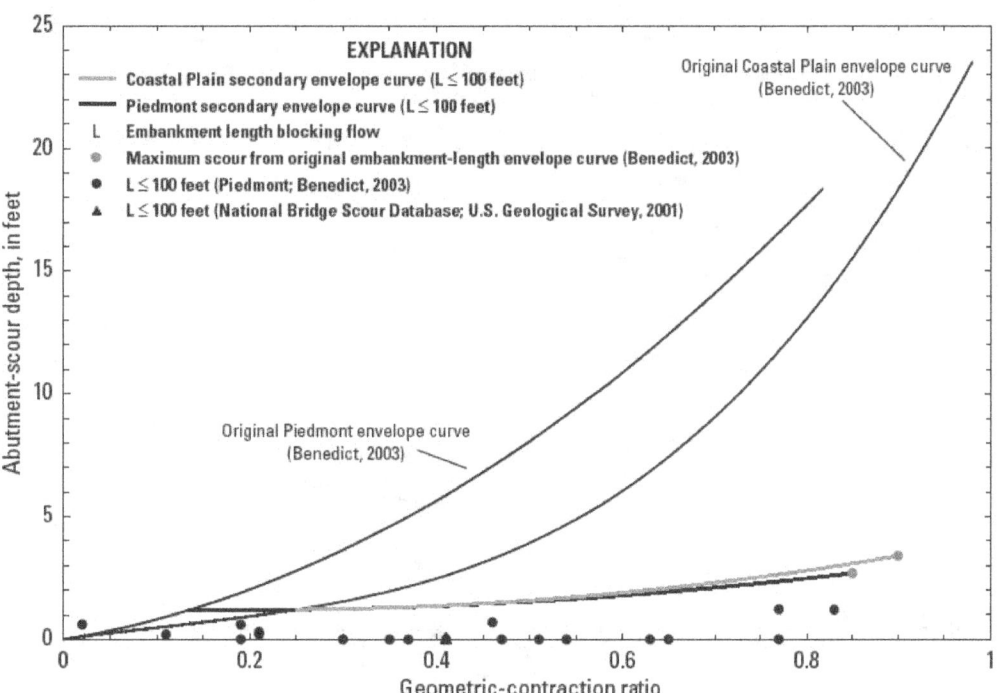

Figure 17.   Relation of clear-water abutment-scour depth to the geometric-contraction ratio for embankment lengths 100 feet or less for selected data in the Piedmont and Coastal Plain of South Carolina.

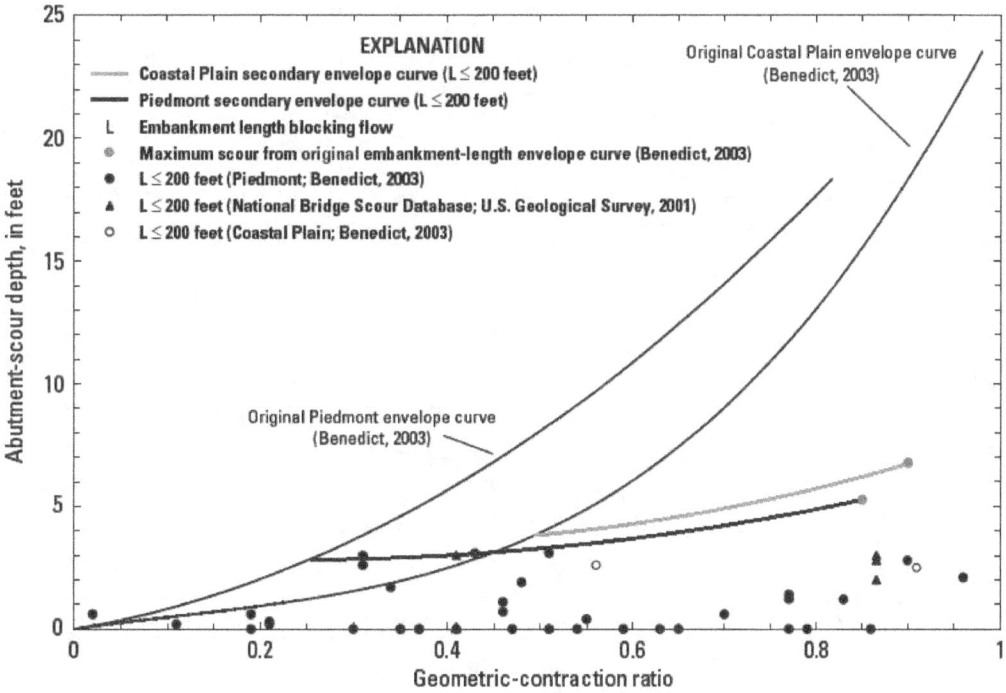

Figure 18.   Relation of clear-water abutment-scour depth to the geometric-contraction ratio for embankment lengths 200 feet or less for selected data in the Piedmont and Coastal Plain of South Carolina.

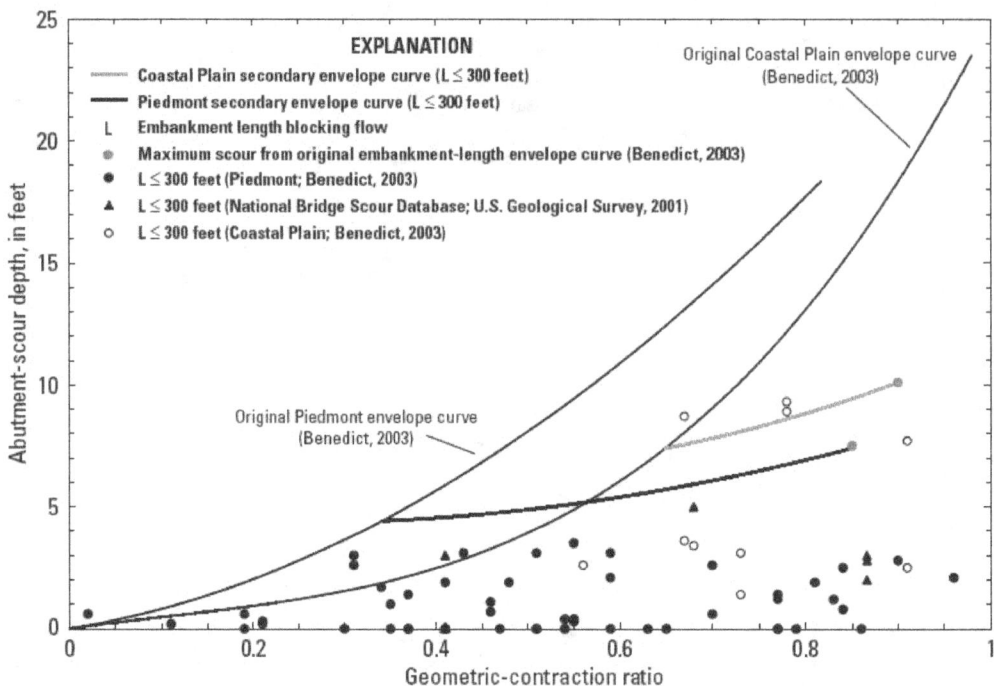

Figure 19.   Relation of clear-water abutment-scour depth to the geometric-contraction ratio for embankment lengths 300 feet or less for selected data in the Piedmont and Coastal Plain of South Carolina.

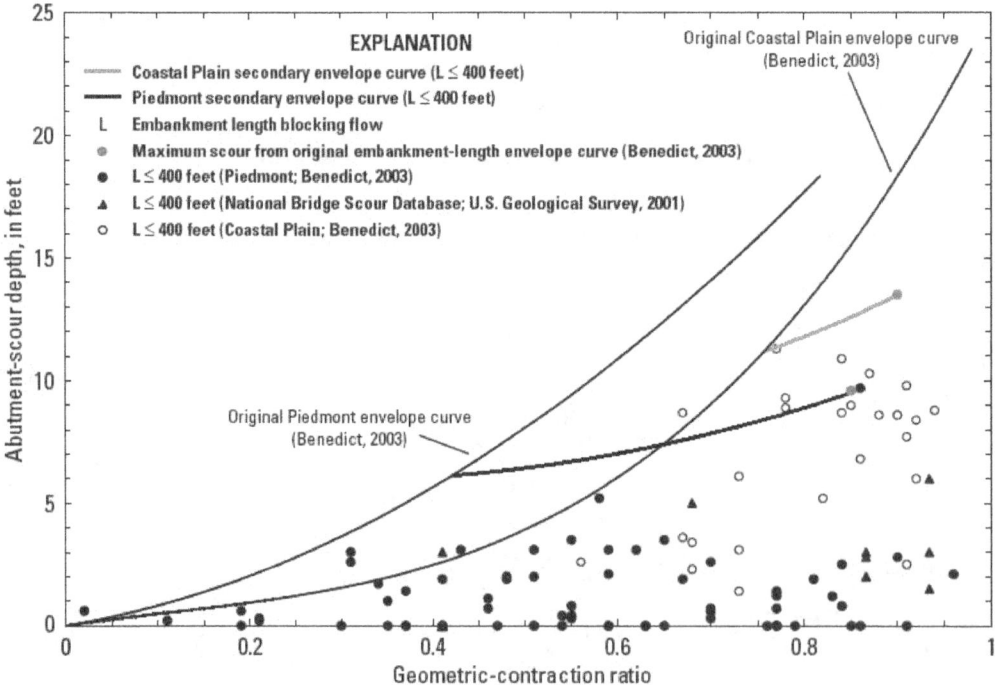

Figure 20.   Relation of clear-water abutment-scour depth to the geometric-contraction ratio for embankment lengths 400 feet or less for selected data in the Piedmont and Coastal Plain of South Carolina.

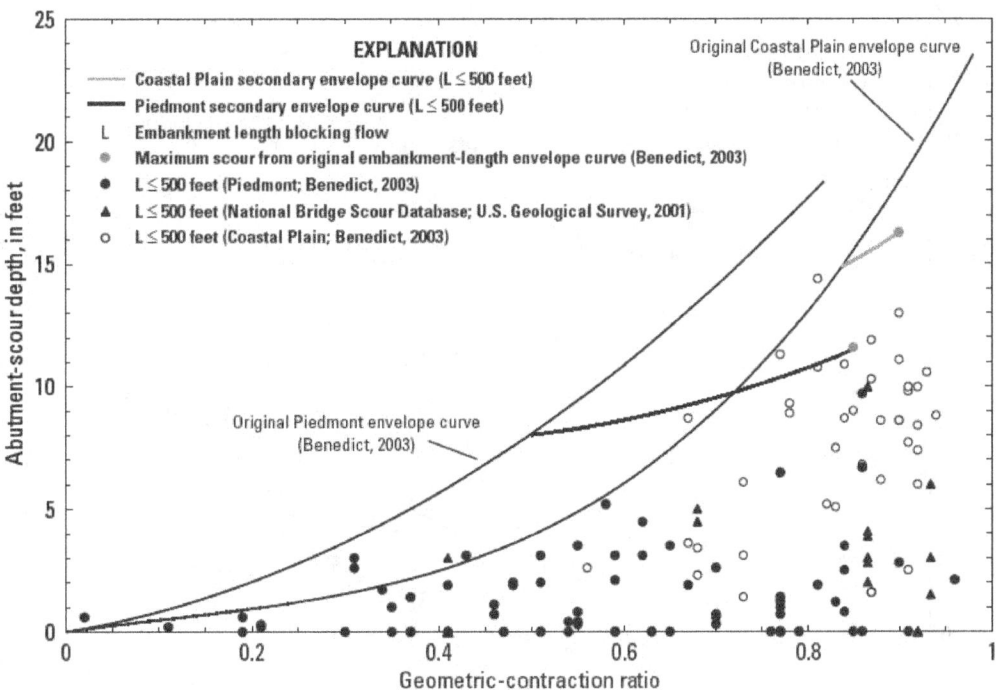

Figure 21. Relation of clear-water abutment-scour depth to the geometric-contraction ratio for embankment lengths 500 feet or less for selected data in the Piedmont and Coastal Plain of South Carolina.

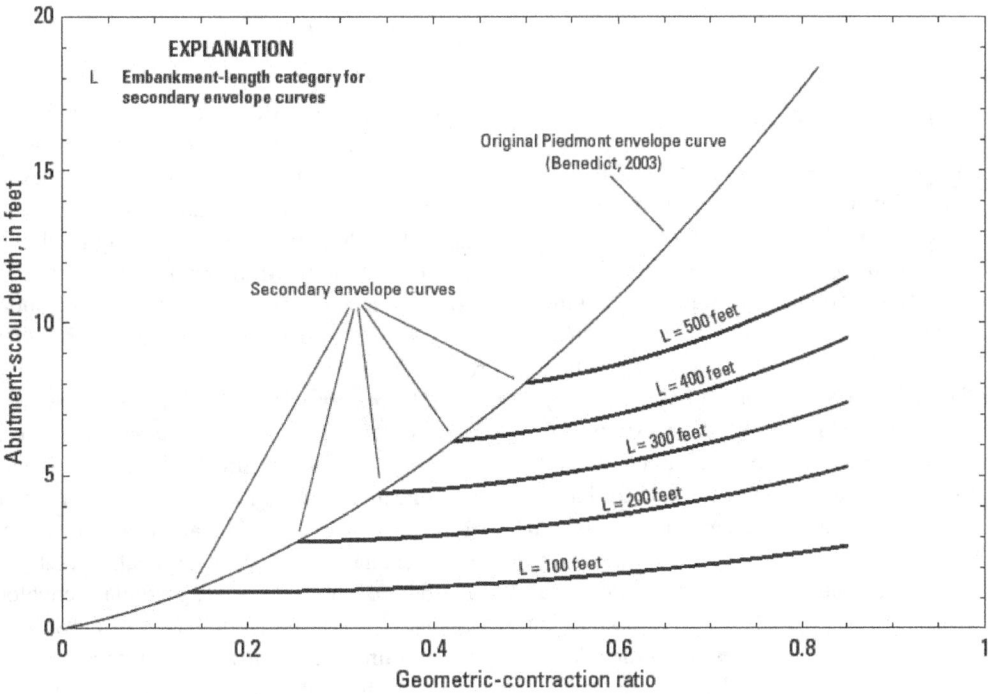

Figure 22. Relation of clear-water abutment-scour depth to the geometric-contraction ratio for selected categories of embankment lengths in the Piedmont of South Carolina.

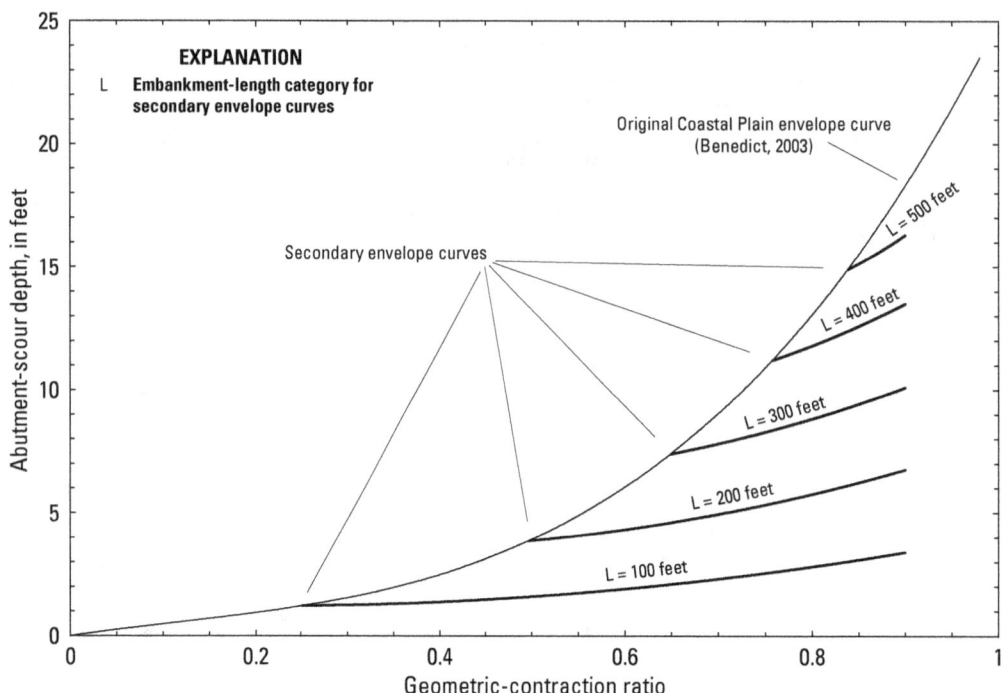

Figure 23. Relation of clear-water abutment-scour depth to the geometric-contraction ratio for selected categories of embankment lengths in the Coastal Plain of South Carolina.

Grouping of Data by Categories

Abutment-scour data for the Piedmont, Coastal Plain, and NBSD were grouped by embankment-length categories of 100-ft increments as follows: 0 to 100 ft, greater than 100 to 200 ft, greater than 200 to 300 ft, greater than 300 to 400 ft, and greater than 400 to 500 ft. The largest embankment length for a given category was used to identify that category. For example, the 200-ft embankment-length category refers to abutment-scour data with embankment lengths greater than 100 ft but less than or equal to 200 ft. The maximum embankment-length category for the Piedmont and Coastal Plain was 500 ft, in part because of the limited data for larger embankment lengths. Additionally, as the embankment-length category exceeds 500 ft, the secondary envelope curves begin to approach and will eventually merge with the original abutment-scour envelope curves (figs. 22 and 23), minimizing any benefit from secondary envelope curves with embankment lengths exceeding 500 ft. Therefore, only data associated with embankment lengths 500 ft or less were utilized in the development of the secondary envelope curves. The smaller dataset from the NBSD was an exception to this. Six of the NBSD abutment-scour data had embankment lengths ranging from 515 to 546 ft. Because of the limited number of NBSD data, these measurements were included and grouped with the embankment-length category of 500 ft. These six measurements fit well within the respective secondary envelope curve and, although their embankment lengths slightly exceed the range of the 500-ft embankment-length category, they provided limited confirmation that the secondary envelope curve for the embankment-length category of 500 ft is reasonable.

The majority of the South Carolina data with embankment lengths of 500 ft or less had drainage areas less than 450 mi^2. Therefore, seven measurements in the Coastal Plain, with drainage areas substantially beyond the range of the majority of the data (between 1,200 and 8,830 mi^2), were excluded from the analysis. [*Note:* These data actually fell within the boundary of the appropriate secondary envelope curves, but because of the limited number of data available for these larger drainage areas it was considered prudent to exclude them.] Additionally, two measurements in the Piedmont and one in the Coastal Plain were excluded from the analysis because of unusual site characteristics, such as a pier or a channel bend that may have affected the measured abutment-scour depth. The data used to develop the secondary envelope

Table 2.   Range of selected site characteristics for field measurements of abutment scour used to develop the modified clear-water abutment-scour envelope curves.

[mi², square mile; ft/ft, foot per foot; ft/s, foot per second; ft, foot; mm, millimeter; <, less than; —, missing data]

Range value	Drainage area (mi²)	Channel slope (ft/ft)	Average blocked approach velocity (ft/s)	Average blocked approach depth (ft)	Embankment length blocking flow (ft)	Geometric-contraction ratio	Median grain size (mm)	Measured abutment-scour depth (ft)
South Carolina Piedmont (Benedict, 2003) (74 measurements)								
Minimum	11	0.00037	0.14	1.0	18	0.02	< 0.062	0.0
Median	76	0.0012	0.92	5.1	208	0.59	0.095	0.7
Maximum	395	0.0029	3.16	13.8	497	0.96	0.99	9.7
South Carolina Coastal Plain (Benedict, 2003) (39 measurements)								
Minimum	6	0.0002	0.18	1.5	127	0.56	< 0.062	1.4
Median	43	0.00076	0.51	4.5	374	0.86	0.21	8.6
Maximum	426	0.0024	1.57	7.1	489	0.94	0.78	14.4
National Bridge Scour Database (U.S. Geological Survey, 2001) (16 measurements)								
Minimum	836	0.00015	—	—	8	0.41	0.001[a]	0.0
Median	845	0.0006	—	—	396	0.87	0.15[a]	3.0
Maximum	1,963	0.0046	—	—	546	0.93	35.0[a]	10.0

[a]Data are missing for 7 measurements.

curves included 74 measurements in the Piedmont and 39 in the Coastal Plain. A summary of the data with the median and range of selected site characteristics is presented in table 2. The NBSD data used to provide limited verification of the South Carolina secondary envelope curves consist of 16 measurements, as summarized in table 2. The drainage areas associated with the NBSD data are outside the range of the South Carolina data used to develop the secondary envelope curves. Although not ideal, the NBSD data were included in the study because of the limited availability of abutment-scour field data for comparison with the South Carolina data.

Selecting Boundaries for the Family of Curves

The original South Carolina clear-water abutment-scour envelope curves for geometric-contraction ratio (Benedict, 2003; figs. 4 and 5) were used as the base graph on which the secondary envelope curves associated with the selected embankment-length categories were superimposed. The original envelope curves (figs. 4 and 5) were used to help define the left boundary for the secondary envelope curves (figs. 17–21). A review of the Piedmont and Coastal Plain data (figs. 4 and 5) indicates that data are limited

for geometric-contraction ratios greater than 0.85 and 0.9, respectively. Therefore, these geometric-contraction ratio values were selected as terminal values for the right boundary of the family of curves for the Piedmont and Coastal Plain, respectively. In the conceptual model (fig. 16), the right boundary for the secondary envelope curves also is associated with the maximum value of abutment-scour depth for each individual curve. To consistently define the maximum abutment-scour depth at the right boundary for each of the secondary envelope curves, the original upper bound envelope curves for embankment length (Benedict, 2003; figs. 2 and 3) were utilized. For example, the embankment-length category of 0 to 100 ft for the Piedmont would have a maximum value of abutment-scour depth of 2.7 ft for the embankment length of 100 ft as determined from figure 2. Therefore, the secondary envelope curve for this embankment-length category would have an upper right boundary located at a geometric-contraction ratio of 0.85 with an abutment-scour depth of 2.7 ft. Defining the upper right boundary of the secondary envelope curves in this manner provides an anchoring point for assisting in drawing each curve. This anchoring point is defined as a red data point in figures 17–21.

Developing Family of Curves

The abutment-scour data for a particular embankment-length category were superimposed on the original envelope curve (fig. 4 or 5) and were reviewed to assess an appropriate secondary envelope curve that would encompass the data for that particular category. The general shape of the secondary envelope curves were developed to conform to the conceptual model (figs. 15 and 16) with the anchoring point (previously described) defining the upper right boundary for each secondary envelope curve. The secondary envelope curves were drawn by hand to encompass the field data associated with the embankment-length category. Additionally, the position and shape of each secondary envelope curve, with respect to its neighboring curves, was considered to ensure that the family of curves was drawn to provide reasonable transitions from curve to curve. On one occasion (fig. 19), data were allowed to slightly exceed the secondary envelope curve, because the data were associated with site conditions that tended to produce larger than normal scour depths. Graphs showing the Piedmont, Coastal Plain, and NBSD data for the embankment-length categories of 100 to 500 ft, along with the secondary envelope curves for those categories are shown in figures 17–21. Summary comments on these secondary envelope curves follow.

Secondary Envelope Curve for Embankment Length of 100 Feet

The Piedmont abutment-scour data, in conjunction with the right-bound anchoring point (described previously), provided a sufficient means to define the secondary envelope curve for the 100-ft embankment-length category in the Piedmont (fig. 17). No Coastal Plain abutment-scour data are available for this embankment-length category. [*Note:* The typically wide Coastal Plain floodplains seldom have small embankment lengths.] Therefore, the Coastal Plain secondary envelope curve for the 100-ft embankment-length category was defined by using judgment in conjunction with the anchoring point and the general trend of the Piedmont secondary envelope curve. The one NBSD measurement included in this category falls within the Piedmont and Coastal Plain secondary envelope curves, providing limited support for the validity of these curves.

Secondary Envelope Curve for Embankment Length of 200 Feet

The Piedmont abutment-scour data, in conjunction with the right-bound anchoring point (described previously), provided a sufficient means to define the secondary envelope curve for the 200-ft embankment-length category in the Piedmont region (fig. 18). The Coastal Plain secondary envelope curve for the 200-ft embankment-length category was defined by utilizing the anchoring point, the two Coastal Plain abutment-scour data included in this category, along with the trend of the Piedmont secondary envelope curve. The NBSD data included in this category fall within the Piedmont and Coastal Plain secondary envelope curves, providing limited support for the validity of these curves.

Secondary Envelope Curve for Embankment Length of 300 Feet

The Piedmont abutment-scour data, in conjunction with the right-bound anchoring point (described previously), provided a sufficient means to define the secondary envelope curve for the 300-ft embankment-length category in the Piedmont (fig. 19). Similarly, the Coastal Plain abutment-scour data and right-bound anchoring point were used to define the secondary envelope curve for the Coastal Plain. There were three Coastal Plain data points that slightly exceeded the secondary envelope curve with the exceedance ranging from 0.4 to 1.2 ft. These data are associated with multiple bridge crossings that make it difficult to define the embankment length and geometric-contraction ratio, and this, in part, is possibly the cause of the discrepancy. However, the exceedance is not excessive and the envelope curve is considered reasonable. The NBSD data included in this category fall within the Piedmont and Coastal Plain secondary envelope curves providing limited support for the validity of these curves.

Secondary Envelope Curves for Embankment Lengths of 400 and 500 Feet

The Piedmont abutment-scour data, in conjunction with the right-bound anchoring points (described previously), provided a sufficient means to define the secondary envelope curves for the 400- and 500-ft embankment-length categories in the Piedmont (figs. 20, 21). Similarly, the Coastal Plain abutment-scour data and right-bound anchoring points were used to define the secondary envelope curves for the Coastal Plain. The NBSD data for these categories fall within the secondary envelope curves, providing limited support for the validity of these curves.

One Piedmont data point located near the anchoring point for the Piedmont 400-ft embankment-length secondary envelope curve (fig. 20) has an embankment length of 408 ft. Although this measurement technically should have been grouped with the embankment-length category of 500 ft, it was purposely grouped with the 400-ft category, because it was close to the category breakpoint of 400 ft and because it provided support for the trend observed in the Piedmont 400-ft secondary envelope curve.

Family of Curves for the Piedmont and Coastal Plain

The family of curves developed for clear-water abutment-scour depth in the Piedmont and Coastal Plain of South Carolina are shown in figures 22 and 23, respectively. The secondary envelope curves were hand drawn to encompass most of the data for a given embankment-length category (figs. 17–21), and to ensure that the position of each secondary envelope curve, with respect to its neighboring curves, provided a reasonable transition from curve to curve. Equations for the family of curves are shown in table 3.

Table 3. Equations for the modified clear-water abutment-scour envelope curves in the Piedmont and Coastal Plain of South Carolina.

[Note: If the geometric-contraction ratio for a given bridge is less than the minimum value, then use the minimum value in equation. ft, foot; ≤, less than or equal to; L, embankment length; y_s, scour depth, in feet; m, geometric-contraction ratio; <, less than]

Embankment-length category	Equation	Limits of the geometric-contraction ratio
Piedmont		
0 ft ≤ L ≤ 100 ft	$y_s = 3.27\,m^2 - 1.12\,m + 1.29$	$0.136 \leq m \leq 0.85$
100 ft < L ≤ 200 ft	$y_s = 6.27\,m^2 - 2.83\,m + 3.16$	$0.255 \leq m \leq 0.85$
200 ft < L ≤ 300 ft	$y_s = 8.33\,m^2 - 4.06\,m + 4.84$	$0.343 \leq m \leq 0.85$
300 ft < L ≤ 400 ft	$y_s = 11.54\,m^2 - 6.78\,m + 6.93$	$0.423 \leq m \leq 0.85$
400 ft < L ≤ 500 ft	$y_s = 15.38\,m^2 - 10.83\,m + 9.61$	$0.503 \leq m \leq 0.85$
Coastal Plain		
0 ft ≤ L ≤ 100 ft	$y_s = 4.64\,m^2 - 1.99\,m + 1.43$	$0.252 \leq m \leq 0.9$
100 ft < L ≤ 200 ft	$y_s = 9.12\,m^2 - 5.55\,m + 4.37$	$0.496 \leq m \leq 0.9$
200 ft < L ≤ 300 ft	$y_s = 13.14\,m^2 - 9.57\,m + 8.07$	$0.649 \leq m \leq 0.9$
300 ft < L ≤ 400 ft	$y_s = 21.30\,m^2 - 19.22\,m + 13.54$	$0.757 \leq m \leq 0.9$
400 ft < L ≤ 500 ft	$y_s = 57.60\,m^2 - 77.53\,m + 39.42$	$0.837 \leq m \leq 0.9$

Guidance and Limitations for Applying the Modified Abutment-Scour Envelope Curves

The modified South Carolina clear-water abutment-scour envelope curves (figs. 22 and 23; table 3) can be useful supplementary tools for assessing clear-water abutment-scour depth at bridges in South Carolina. When using the modified envelope curves to assess scour potential, one must select a reference surface, estimate the embankment length and the geometric-contraction ratio, select the appropriate abutment-scour envelope curve, evaluate other scour components in the abutment region, and consider the limitations associated with the envelope curves.

1. Select a reference surface:

The reference surface should be the average, undisturbed floodplain elevation in the abutment-scour region and this can be determined by reviewing floodplain elevations from SCDOT road plans, surveyed cross sections, and (or) site visit observations. For additional details refer to Benedict (2003).

2. Estimate of the embankment length and the geometric-contraction ratio:

The modified clear-water abutment-scour envelope curves can be sensitive to the selection of embankment length and the geometric-contraction ratio. Therefore, it is important that good

estimates of these variables be obtained. These variables can be determined by using flow models, topographic maps, and road plans, and when possible, all three sources should be used for verification. When discrepancies exist between these sources, judgment should be used to determine the most reasonable estimate of the explanatory variables.

3. Select the appropriate abutment-scour envelope curve:

Clear-water abutment-scour depth can be assessed using the original envelope curves (Benedict 2003; figs. 2–5) or the modified envelope curves (figs. 22 and 23; table 3). Criteria for selecting the appropriate envelope curve will be based on the regional location and site characteristics for the bridge of interest. To utilize the modified clear-water abutment-scour envelope curves, the embankment lengths must be 500 ft or less and the geometric-contraction ratio should not exceed 0.85 or 0.9 for the Piedmont or Coastal Plain, respectively. If the geometric-contraction ratio for the bridge of interest is less than the minimum value specified in table 3, then use the minimum value. Additionally, the site characteristics for the bridge of interest should fall within the range of the appropriate South Carolina regional data used to develop the modified envelop curves as shown in table 2. If these criteria are not met, then the original clear-water abutment-scour envelope curves must be used (figs. 2–5) following the application guidance from Benedict (2003).

For embankment lengths that fall between the embankment-length categories of the modified envelope curves (figs. 22 and 23; table 3), it is possible to interpolate between the secondary envelope-curves in order to refine the estimate of clear-water abutment-scour depth. However, it may be prudent to apply the modified envelope curves only by embankment-length category. Using this category application, there may be cases when the modified envelope curves provide larger values of clear-water abutment-scour potential in comparison to the original clear-water abutment-scour envelope curves. In such cases, it is recommended that the smallest value of clear-water abutment-scour potential be used. Therefore, if the objective is to determine a refined estimate of clear-water abutment-scour potential, it is recommended that both the original envelope curves (Benedict, 2003) and the modified envelope curves be applied, with the smallest value of clear-water abutment-scour potential being used as the refined estimate.

4. Contraction and pier scour within the abutment-scour region:

Benedict (2003) concluded that contraction scour should not be considered a contributing component to total scour in the abutment region. With respect to pier scour, Benedict (2003) concluded that piers located in the abutment-scour region, having widths 2.3 ft or less, have negligible influence on total scour depth. Therefore, when using the modified clear-water abutment-scour envelope curves to assess total scour depth at abutments, no adjustment is required for contraction scour or pier scour with pier width 2.3 ft or less. When the pier width exceeds 2.3 ft, judgment should be used to determine if the results of the clear-water abutment-scour envelope curves should be adjusted to account for the effects of the wider piers. Judgment also should be used if the effects of debris and (or) pier skew must be considered. The above guidance for piers is slightly modified for the Piedmont. When the range of clear-water abutment-scour depth in the Piedmont is estimated to be 5 ft or less with the modified envelope curves, judgment should be used to account for the effect of pier scour within the abutment region. For additional guidance refer to Benedict (2003).

5. Bridges 240 feet or less:

Benedict (2003) notes that bridge lengths of 240 ft or less tend to form a large, single scour hole that encompasses most of the bridge opening rather than separate left and right abutment scour holes. When applying the modified clear-water abutment-scour envelope curves at such bridges, use the longest of the left or right embankment length to assess the clear-water abutment-scour potential and, assume that the estimated scour depth will represent the depth of the large, single scour hole that will likely extend from abutment toe to abutment toe. For additional information regarding this type of bridge, refer to Benedict (2003).

6. Limitations of the modified abutment-scour envelope curves:

The modified South Carolina clear-water abutment-scour envelope curves (figs. 22 and 23; table 3) can be useful supplementary tools for assessing clear-water abutment-scour potential at bridges in South Carolina. However, the following limitations of these empirical envelope curves should be kept in mind when evaluating the potential for clear-water abutment scour in South Carolina.

- The modified envelope curves were developed from a limited sample of bridges in the Piedmont and Coastal Plain, and it is possible that scour depths could exceed the envelope curves. Therefore, applying a safety factor to the modified envelope curves may be prudent.

- Application of the modified envelope curves should be limited to bridges having site characteristics similar to the South Carolina data used to develop the envelope curves (table 2).

- The modified envelope curves were developed using field data representing scour resulting from flows near the 1-percent AEP (100-year) flow and should not be used to evaluate clear-water abutment-scour depths for larger flows, such as the 0.2-percent AEP (500-year) flow.

- The washout of road embankments is a frequent abutment-scour problem in South Carolina. Typically, washout occurs at shorter bridges that produce a large contraction of flow. The modified envelope curves cannot be used to assess this type of abutment scour.

- The modified envelope curves do not account for adverse field conditions that may affect abutment scour at bridges, such as channel bends and debris.

- A limited number of measurements (12) associated with multiple-bridge crossings were used in the development of the modified clear-water abutment-scour envelope curves. On occasion these data exceeded the secondary envelope curves (fig. 19). Therefore, it is not recommended that the modified envelope curves be applied to multiple bridges. Instead, guidance from Benedict (2003) should be used for assessing abutment-scour potential at multiple-bridge crossings.

Although the modified South Carolina clear-water abutment-scour envelope curves presented in this report can serve as a valuable supplementary tool in assessing clear-water abutment-scour depths in South Carolina, the noted limitations restrict their use. Therefore, the modified envelope curves should not be relied upon as the only tool for assessing clear-water abutment-scour potential. To best assess potential scour, one should compile and study the available information for a given site and then bring sound engineering principles to bear on the final estimate of potential abutment-scour depth.

The Modified South Carolina Live-Bed Contraction-Scour Envelope Curve

In contrast to clear-water abutment scour that typically occurs on the stream floodplain at bridges in South Carolina, live-bed contraction scour occurs in the main channel. Research indicates that contraction-scour depth will be strongly correlated to the degree of contraction produced by a bridge (Laursen, 1960; Melville and Coleman, 2000; Mueller and Wagner, 2005), which can be expressed as the geometric-contraction ratio (see previous definition). Based on this trend, Benedict and Caldwell (2009) developed a live-bed contraction-scour envelope curve that displayed the range and trend of the upper bound of historic live-bed contraction scour in South Carolina, using the geometric-contraction ratio as the explanatory variable (fig. 24). The three measurements that exceed the envelope curve are associated with site conditions (channel bends or debris) that will tend to increase the potential for scour. Although the live-bed contraction-scour envelope curve has limitations (Benedict and Caldwell, 2009), it can be used as a supplementary tool to evaluate predicted live-bed contraction scour as well as the potential for live-bed contraction scour in South Carolina.

An objective of the current (2011) investigation is to modify the live-bed contraction-scour envelope curve (Benedict and Caldwell, 2009; fig. 24) that utilized a single explanatory variable to include a family of curves that utilized two explanatory variables, thus providing a means to refine the assessment of the upper bound of historic live-bed contraction scour in South Carolina. The following report sections document the development of the modified live-bed contraction-scour envelope curve with particular focus on (1) the field data used in the analysis, (2) the conceptual model, (3) procedures used to develop the modified envelope curves, and (4) guidance for applying the modified envelope curve.

Selected Field Data Used in the Analysis

In an extensive literature review by Wagner and others (2006), only 7 measurements of live-bed contraction scour, which contained adequate supporting data, could be identified for use in assessing scour-prediction equations. This highlights the dearth of field measurements for live-bed contraction scour and the need to promote field investigations to collect additional data. Because of the limited field measurements of live-bed contraction scour, the USGS conducted several field investigations of contraction scour (Mueller and Wagner, 2005; Wagner and others, 2006; Benedict and Caldwell, 2009) with the objective to form databases that could be used to evaluate the performance of scour-prediction equations and understand contraction-scour trends within the field setting. The USGS live-bed contraction scour field data collected in

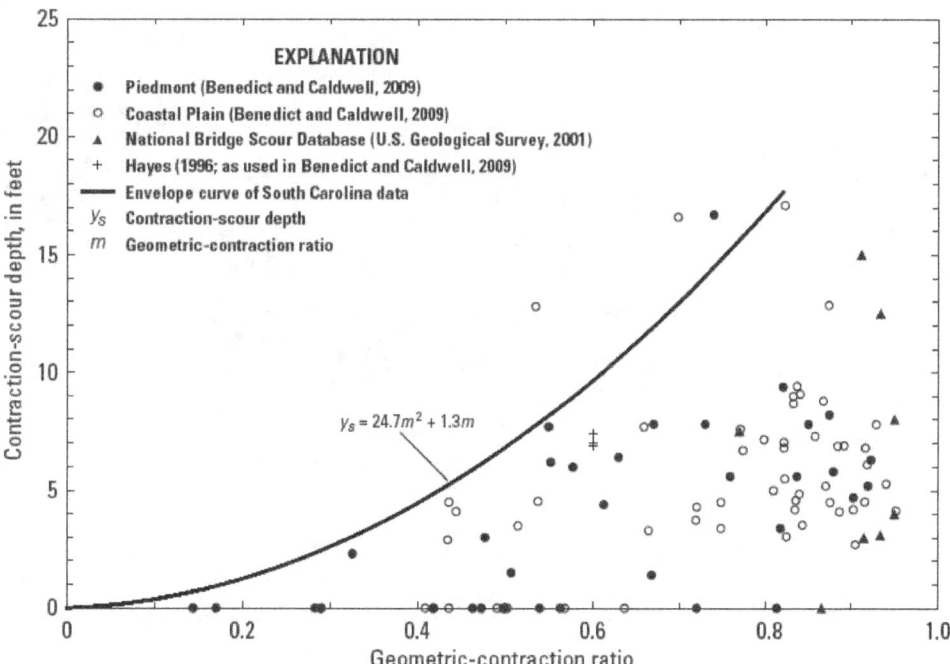

Figure 24. Relation of measured live-bed contraction-scour depth to the geometric-contraction ratio for selected data in the Piedmont and Coastal Plain of South Carolina (modified from Benedict and Caldwell, 2009).

these investigations included 89 measurements from South Carolina (Benedict and Caldwell, 2009) and 28 measurements from the USGS National Bridge Scour Database (NBSD; U.S. Geological Survey, 2001; Wagner and others, 2006). The South Carolina database was the primary source for developing the original (Benedict and Caldwell, 2009; fig. 24) South Carolina live-bed contraction-scour envelope curve and also served as the primary data source for this investigation. The NBSD data were used for limited confirmation of the patterns of the South Carolina data. Additionally, 42 measurements of clear-water contraction scour from South Carolina

(Benedict, 2003) associated with short bridges (240 ft or less) with severe contractions were used to help verify the patterns in the live-bed contraction-scour measurements. A brief description of these data and their limitations follows.

South Carolina Live-Bed Contraction-Scour Data

The South Carolina bridge-scour data include 89 measurements of historic live-bed contraction-scour collected at 89 bridges in the Piedmont and Coastal Plain of South Carolina (Benedict and Caldwell, 2009; fig. 25; table 4). These scour

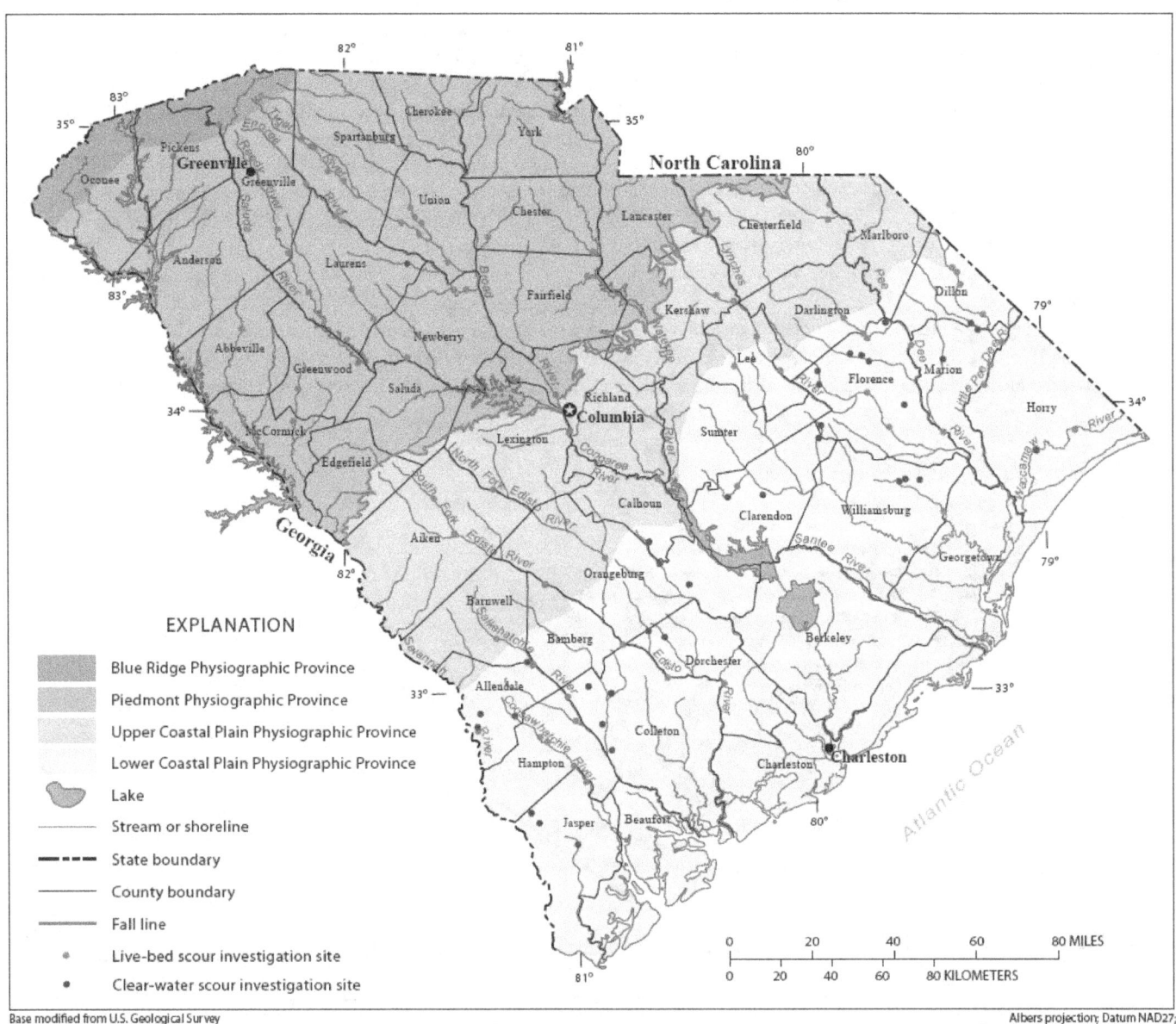

Figure 25.   Location of selected live-bed and clear-water contraction-scour investigation sites in South Carolina (modified from Benedict and Caldwell, 2009).

Table 4. Range of selected site characteristics for field measurements of contraction scour.

[mi², square mile; ft/ft, foot per foot; ft/s, feet per second; ft, feet; mm, millimeter]

Range value	Drainage area (mi²)	Channel slope (ft/ft)	Average approach channel velocity (ft/s)	Average approach channel depth (ft)	Average channel width (ft)	Geometric-contraction ratio	Median grain size (mm)	Measured contraction-scour depth (ft)
South Carolina Piedmont (Benedict and Caldwell, 2009) *(35 measurements)*								
Minimum	21	0.00015	2.4	7.7	41	0.14	0.51	0.0
Median	148	0.001	5.6	15.7	87	0.61	0.78	3.4
Maximum	5,250[a]	0.0021	11.6	28.3	788	0.92	1.7	16.7
South Carolina Coastal Plain (Benedict and Caldwell, 2009) *(54 measurements)*								
Minimum	17.2	0.00007	1.1	4.7	21	0.29	0.18	0.0
Median	521	0.00031	2.7	12.5	93	0.82	0.59	4.6
Maximum	9,360[b]	0.002	7.1	39.0	785	0.95	1.7	17.1
South Carolina Clear-Water Contraction Scour (Benedict, 2003) *(42 measurements)*								
Minimum	6.1	0.00015	0.05	2.0[d]	floodplain	0.77	0.06	0.9
Median	32.2	0.001	0.5	4.3[d]	floodplain	0.91	0.24	9.6
Maximum	8,230[c]	0.0021	0.9	11.7[d]	floodplain	0.98	0.78	23.6
National Bridge Scour Database (U.S. Geological Survey, 2001) *(9 measurements)*								
Minimum	10.3	0.00006	0.7[e]	7.9[e]	42[e]	0.77	0.1[e]	0.0
Median	845	0.0005	3.4[e]	16.0[e]	90[e]	0.92	0.2[e]	4.0
Maximum	16,010	0.001	5.2[e]	43.0[e]	300[e]	0.95	1.6[e]	15.0

[a]About 94 percent of the study sites in the Piedmont have drainage areas less than 760 mi².

[b]About 80 percent of the study sites in the Coastal Plain have drainage areas less than 1,860 mi².

[c]About 95 percent of the study sites in the South Carolina clear-water data have drainage areas less than 265 mi².

[d]These sites are associated with swampy floodplains with shallow, poorly defined channels; the average approach flow depth was based on the average floodplain depth.

[e]Data are not available at all sites.

depths range from 0 to 17.1 ft and are considered to represent the maximum live-bed contraction-scour depth that has occurred at the bridge since construction. Live-bed contraction-scour holes in South Carolina occur in the main channel of streams and are typically inundated and partially or totally refilled with sediments, making the measurement of these scour holes problematic. Therefore, Benedict and Caldwell (2009) utilized a ground-penetrating radar (GPR) system to measure historic live-bed contraction scour at the selected bridges. (See Benedict and Caldwell (2009) for details regarding the application and limitations of GPR for the investigation.) A grab sample of sediment in the main channel was obtained

at each site, and because sediment characteristics can vary spatially in the field, these samples may not fully represent the sediment characteristics at a given site.

As with the South Carolina clear-water-abutment scour data, the South Carolina live-bed contraction-scour data were measured during low flows and a one-dimensional step-backwater model, WSPRO (Shearman, 1990), was used to estimate the hydraulic conditions that may have produced the observed scour. Historic flood records were available at or near 68 of the 89 bridges, and these were used to estimate the historic peak flows at those sites. Historic peak flows ranged from 0.34 to 2.85 times the 1-percent AEP (100-year) flow with a median value of 0.95. The historic peak flows were used in the step-backwater models to estimate the hydraulic characteristics at these sites. For the remaining 21 bridges, a risk analysis associated with the bridge age indicated that these sites likely had experienced flows equaling or exceeding 70 percent of the 1-percent AEP (100-year) flow. Therefore, the 1-percent AEP flow was used in the step-backwater models at these sites.

The hydraulic characteristics approximated with the model may introduce error into the analysis of the live-bed contraction-scour data, making such analysis less than ideal. However, the large number of field measurements (89) in the database should provide understanding into regional patterns for live-bed contraction scour in South Carolina. Because of regional distinctions, the live-bed contraction-scour data from the Piedmont and Coastal Plain were grouped into separate databases. Table 4 lists the median and range of selected site characteristics for the data and provides some understanding of the regional differences. For additional details regarding the South Carolina live-bed contraction-scour data, refer to Benedict and Caldwell (2009).

National Bridge Scour Database

The NBSD (U.S. Geological Survey, 2001) contains 44 measurements of contraction scour, 27 that are live-bed contraction scour, which were taken at various bridge sites throughout the United States. A review of the NBSD live-bed contraction-scour data concluded that 18 of the measurements should be excluded from the current investigation (2011) for various reasons, such as debris accumulation, flows associated with the measured scour that were less than a 20-percent AEP (5-year) flow, time-series measurement duplicates of other measurements in the series, and insufficient supporting data. The remaining nine measurements were used for comparison and limited confirmation of the patterns of the original South Carolina live-bed contraction-scour envelope curve (Benedict and Caldwell, 2009; fig. 24). The modification of the original live-bed contraction-scour envelope curve focused primarily on the development of secondary envelope curves for smaller drainage areas. A review of the nine selected NBSD data measurements showed that seven of them were associated with sites having drainage areas greater than 800 mi^2, which was beyond the limits of the drainage-area categories used to develop the secondary envelope curves. Therefore, only two NBSD data measurements could be used to compare with the patterns of the modified South Carolina live-bed contraction-scour envelope curve. Hydraulic characteristics associated with the NBSD scour measurements were developed from a combination of historic flow records and flow models similar to the methods used for the South Carolina data. Table 4 lists the median and range of selected site characteristics for the nine selected NBSD field data measurements. Additional details associated with the data can be found at the NBSD web page (*http://water.usgs.gov/osw/techniques/bs/BSDMS/index.html*, accessed July 22, 2011) as well as in Wagner and others (2006).

South Carolina Clear-Water Contraction-Scour Data

The South Carolina bridge-scour data include 42 measurements of clear-water contraction scour (Benedict, 2003; table 4; fig. 25) associated with shorter bridges (about 240 ft or less in length) that typically produce large contractions of flow and typically have substantial scour. These sites include floodplain relief bridges and bridges crossing swamps with poorly defined channels. These bridges typically develop a large single scour hole with a top width that typically encompasses the entire bridge opening. The severe contractions associated with these bridges substantially increase flow velocities at the bridge and tend to produce turbulent flow patterns similar to those that cause abutment scour. Because of the similarity of flow patterns to those for abutment scour, Benedict (2003) considered these 42 measurements as special cases of clear-water abutment scour. However, because of the short bridge lengths and the large contractions produced by these bridges, it is also appropriate to consider these data as cases of severe clear-water contraction scour. Drainage areas associated with the 42 measurements range from 6.1 to 8,230 mi^2 (table 4), but data are primarily associated with smaller drainage areas less than 265 mi^2. The clear-water contraction-scour depths for the selected South Carolina data range from 0.9 to 23.6 ft. The selected South Carolina clear-water contraction-scour data have similar median grain sizes (0.06 to 0.78 millimeters) to those of the South Carolina live-bed data. The reference surface for estimating contraction-scour depths at these sites was the thalweg of the shallow, swampy channel running through the bridge. Where this shallow channel did not exist, the reference surface was the average elevation of the floodplain. Most of the selected South Carolina clear-water contraction-scour data likely experienced floods equaling or exceeding 70 percent of the 100-year flow magnitude (Benedict, 2003). Therefore, values for hydraulic characteristics were estimated with the WSPRO (Shearman, 1990) model using the 1-percent AEP (100-year) flow to approximate the hydraulic conditions that may have produced the observed scour. The modification of the original live-bed contraction-scour envelope curve focused primarily on the development of secondary envelope curves for smaller drainage areas (200 mi^2 or less). A review of the 42 measurements of clear-water contraction scour showed that 4 of the measurements were associated with sites having drainage areas greater than 200 mi^2, leaving 38 data measurements that could be used to compare with the patterns of the modified South Carolina live-bed contraction-scour envelope curve.

Because researchers have traditionally separated the analyses of clear-water and live-bed scour, utilizing live-bed and clear-water contraction-scour data in this investigation initially may seem inappropriate. However, if the purpose of this study is to understand the upper bound of contraction scour on a regional basis, some data comparisons are appropriate. Scour will display regional trends that are influenced by the hydrology and geology of a given region (Benedict, 2003; 2007). In South Carolina, the regional geology typically constrains the rate and limits of scour and the equilibrium scour depths obtained in the simplistic setting of the laboratory often cannot be attained in the field. Although the South Carolina clear-water contraction-scour data will have some differences from the live-bed contraction-scour data, both are subject to the same regional influences. Therefore, it is reasonable to assume there will be similarities in these two datasets, which justifies using the clear-water contraction scour data to help validate the trends of the live-bed contraction-scour data.

Conceptual Model for the Modified Live-Bed Contraction-Scour Envelope Curve

The conceptual model used to modify the South Carolina live-bed contraction-scour envelope curve assumes that (1) the geometric-contraction ratio is a strong explanatory variable for live-bed contraction scour and can be utilized to develop envelope curves for the upper bound of historic scour, as demonstrated by Benedict and Caldwell (2009; fig. 24), and (2) the potential for live-bed contraction-scour depth will increase with drainage area, such that the upper bound of scour at bridges associated with smaller drainage areas will be less than that associated with larger drainage areas. These two assumptions can be utilized in a common envelope curve to develop a family of curves. The family of curves would consist of live-bed contraction-scour data grouped into categories of drainage area and plotted against the geometric-contraction ratio. Support for this approach can be substantiated by a brief review of selected laboratory and field data.

Investigations of contraction scour indicate that the degree of contraction produced by a bridge will be a strong explanatory variable for contraction-scour depth. This is highlighted by a review of published contraction-scour equations by Melville and Coleman (2000) and Mueller and Wagner (2005) in which all of the reviewed equations included some form of a contraction ratio as a primary explanatory variable. As noted previously, the degree of contraction at a bridge is defined by the geometric-contraction ratio. Laboratory data from Dey and Raikar (2005) show the general trend of contraction-scour depth with respect to this variable (fig. 26). The pattern in figure 26 indicates that the upper bound of contraction-scour depth increases with increasing geometric-contraction ratio. This pattern also is observed in the upper bound of the South Carolina and NBSD data (fig. 24).

A review of the live-bed contraction-scour data from South Carolina (Benedict and Caldwell, 2009; table 4) and the NBSD (U.S. Geological Survey, 2001; table 4), indicates that the upper bound of contraction-scour depth increases with increasing

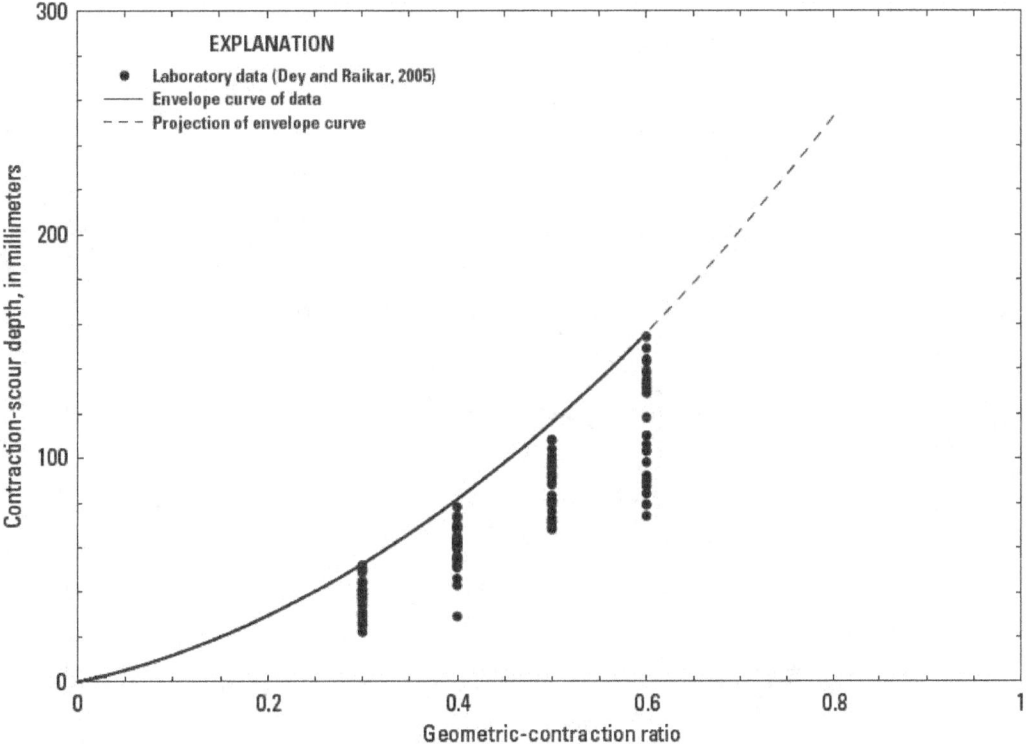

Figure 26.   Relation of contraction-scour depth to the geometric-contraction ratio for selected laboratory data.

drainage area (fig. 27). [*Note:* One outlier associated with substantial debris accumulation can be discarded.] The trend is not as clear in the Piedmont live-bed contraction-scour data. However, the Coastal Plain and NBSD data have similar upper bounds. The South Carolina clear-water contraction-scour data (Benedict, 2003; table 4) associated with short bridges with severe contractions display a similar pattern, which supports the validity of the trend of the live-bed contraction-scour data. The specific reasons why smaller drainage areas have a lower upper bound of scour are not fully understood. However, it is speculated that the shorter flow durations associated with smaller drainage areas may tend to produce smaller contraction-scour depths. Additionally, the thickness of channel alluvium sediments above scour-resistant subsurface materials, such as bedrock, tends to be smaller for smaller drainage areas (fig. 28). [*Note:* See Benedict and Caldwell (2009) for description of the data shown in figure 28.] With the exception of a few outliers in figure 28, this general trend is evident, especially in the Coastal Plain data. Although a definitive explanation for the three outliers in figure 28 is uncertain, it is speculated that unique features associated with these sites (transitions to flatter slopes and (or) wider floodplains) tend to produce deeper deposits of sands, causing them to exceed the majority of the data.

Based on the above-mentioned patterns of the laboratory and field data, it was concluded that the South Carolina live-bed contraction-scour envelope curve could be modified by grouping the South Carolina live-bed contraction-scour data into categories of drainage area, and then plotting them against the geometric-contraction ratio. Application of this conceptual model for the development of a family of curves to modify the South Carolina live-bed contraction-scour envelope curve is described in the following section.

Development of the Modified Live-Bed Contraction-Scour Envelope Curve

To apply the conceptual model to the South Carolina live-bed contraction-scour data, a process similar to the development of the modified clear-water abutment-scour envelope curves was used. Data were grouped into categories of drainage-area ranges, the data for each category were superimposed on the original South Carolina live-bed contraction-scour envelope curve (fig. 24), and judgment was used to construct a family of curves that were reasonably located with respect to each other. A more detailed description of this process follows.

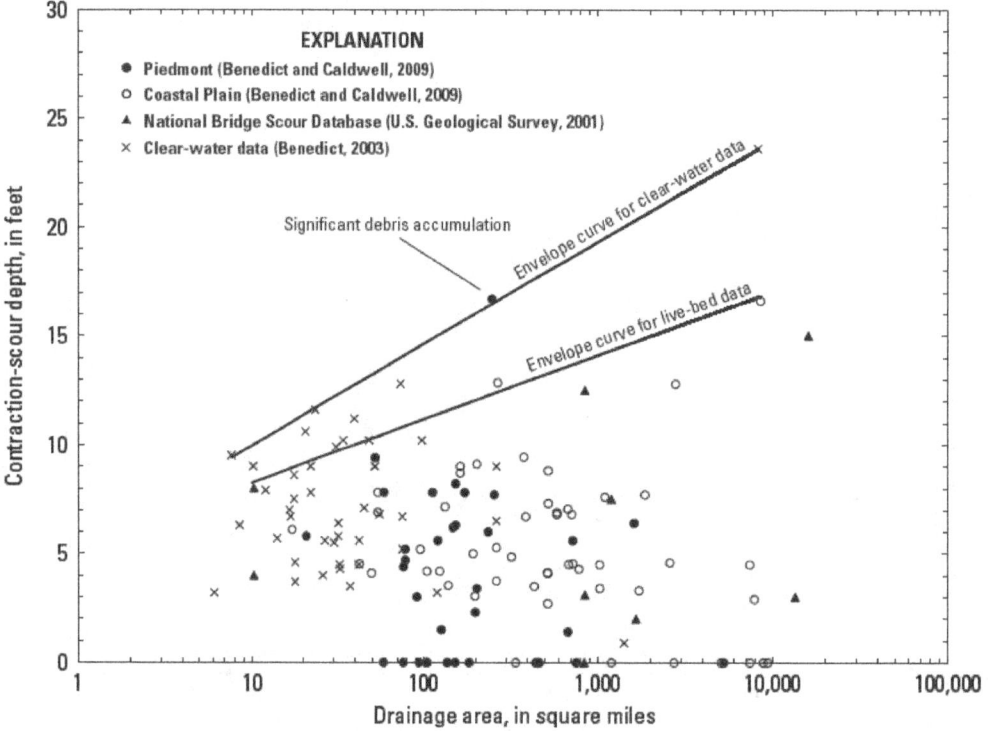

Figure 27.   Relation of contraction-scour depth to drainage area for selected field data.

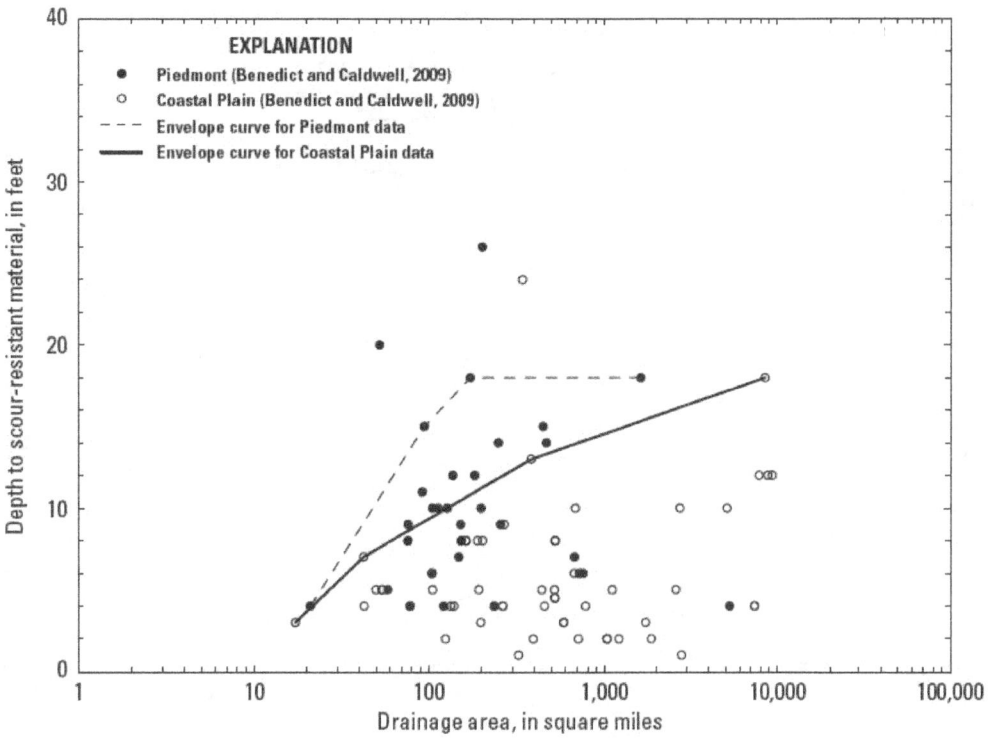

Figure 28. Relation of depth to scour-resistant material in the stream channel to drainage area for selected field data in the Piedmont and Coastal Plain of South Carolina.

Benedict and Caldwell (2009) concluded that the upper bound trends of the Piedmont and Coastal Plain live-bed contraction-scour data were similar enough to develop a single envelope curve that encompassed the data of both regions, rather than separate curves for each region. This approach also was used in developing the secondary envelope curves for the modified live-bed contraction scour family of curves. Figures 29 and 30 show the data and secondary envelope curves developed for each category of drainage area and will be referred to in the following discussion. Following the approach used to display the data for the modified clear-water abutment-scour envelope curves (figs. 17–22), figure 30 displays the data for the 200-mi^2 category, as well as the data from the smaller category of 100 mi^2 as shown in figure 29. The full family of curves is shown in figure 31 and demonstrates how the secondary envelope curves relate to each other.

Grouping of Data by Categories

The live-bed contraction-scour data for the South Carolina and NBSD datasets were grouped according to drainage-area and placed into categories based on 100-mi^2 increments. The maximum drainage-area category was 200 mi^2, in part because of the limited amount of data associated with larger drainage areas. Additionally, the secondary envelope curve for the drainage-area category of 200 mi^2 (fig. 30) is close to the original live-bed contraction-scour envelope curve (Benedict and Caldwell, 2009), minimizing any benefit from secondary envelope curves developed for drainage areas exceeding 200 mi^2. Therefore, only data associated with drainage areas 200 mi^2 or less were utilized in the development of the secondary envelope curves. This upper limit of drainage area constrained the NBSD data to two contraction-scour measurements, substantially limiting the usefulness of that dataset for verifying the patterns of the secondary envelope curves. The South Carolina live-bed contraction-scour data used to develop the secondary envelope curves included 24 measurements in the Piedmont and 15 in the Coastal Plain. A summary of the data with the median and range of selected site characteristics is presented in table 5. The NBSD live-bed contraction-scour data (2 measurements) and the South Carolina clear-water contraction-scour data (38 measurements) were used to provide limited verification of the South Carolina live-bed contraction-scour secondary envelope curves. The median and range of selected site characteristics for these datasets are summarized in table 5.

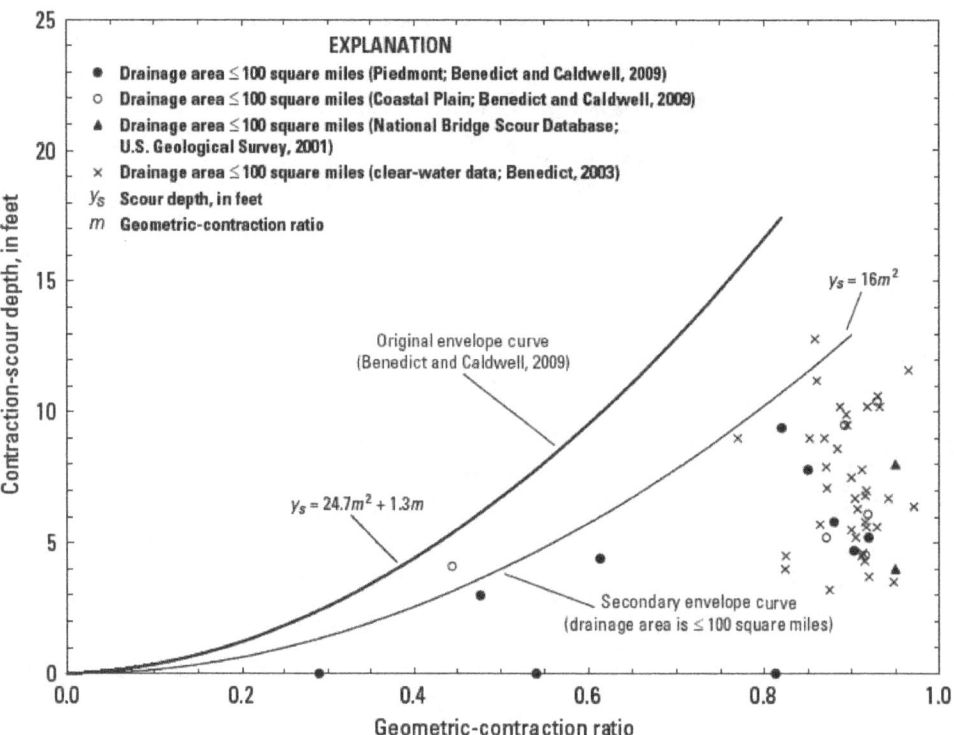

Figure 29. Relation of live-bed contraction-scour depth to the geometric-contraction ratio for drainage areas 100 square miles or less for selected data in the Piedmont and Coastal Plain of South Carolina.

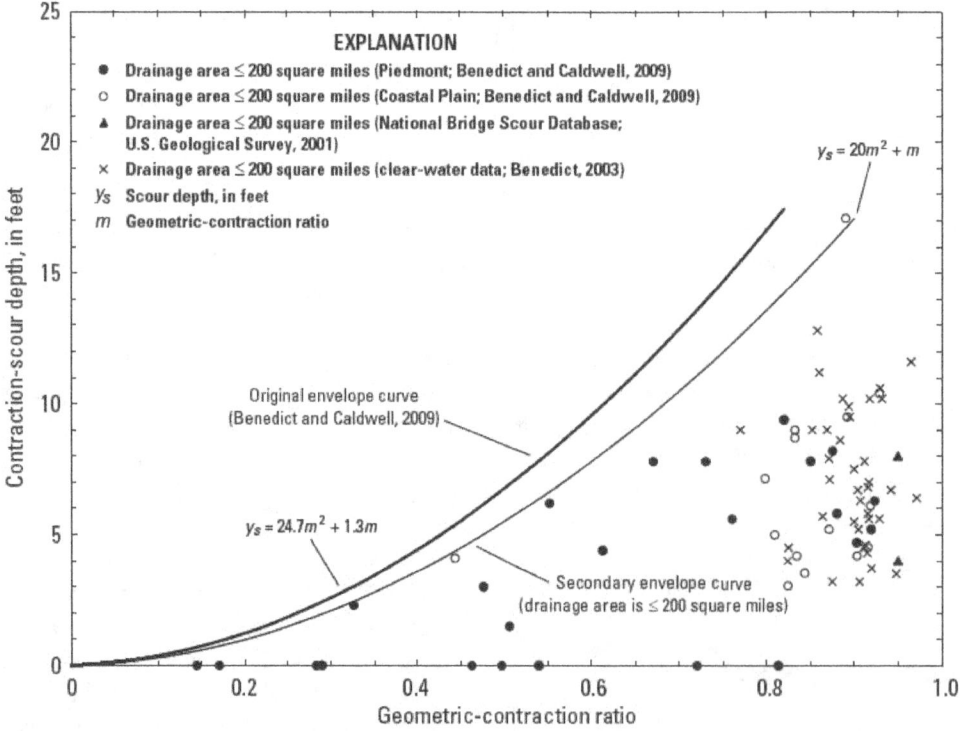

Figure 30. Relation of live-bed contraction-scour depth to the geometric-contraction ratio for drainage areas 200 square miles or less for selected data in the Piedmont and Coastal Plain of South Carolina.

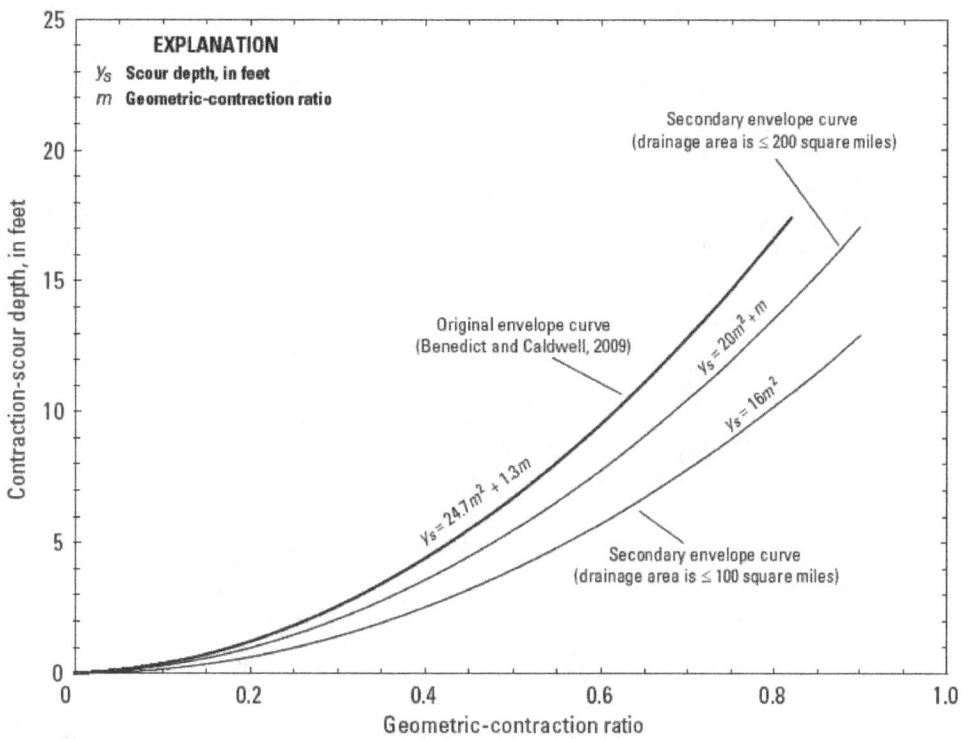

Figure 31.   Relation of live-bed contraction-scour depth to the geometric-contraction ratio for selected categories of drainage area for selected data in the Piedmont and Coastal Plain of South Carolina.

Selecting Boundaries for the Family of Curves

The original South Carolina live-bed contraction-scour envelope curve (Benedict and Caldwell, 2009; fig. 24) was used as the base graph on which the secondary envelope curves associated with the drainage-area categories were superimposed. The original envelope curve (fig. 24) was used to help define the shape of the secondary envelope curves (figs. 29 and 30). Theoretically, a geometric-contraction ratio of zero will produce zero scour. Therefore, the left boundary of the secondary envelope curves was assumed to be at the coordinate location where the geometric contraction ratio and scour depth are both zero. A review of the South Carolina live-bed contraction-scour data (fig. 24) indicates that data are limited beyond the geometric-contraction ratio of 0.90. Therefore, this value was selected as the terminal value for the right boundary for the secondary envelope curves.

Developing Family of Curves

The contraction-scour data for a particular drainage-area category were superimposed on the original live-bed contraction-scour envelope curve (Benedict and Caldwell, 2009; fig. 24) and were reviewed to assess an appropriate secondary envelope curve that would encompass the data for that particular category. The secondary envelope curves

were drawn by hand to encompass the field data associated with each drainage-area category. Additionally, the position and shape of each secondary envelope curve, with respect to its neighboring curves, was considered to ensure that the family of curves was drawn to provide reasonable transitions from curve to curve. Some data points were allowed to slightly exceed the secondary envelope curve (figs. 29 and 30), because the data were associated with site conditions that tended to produce larger than normal scour depths. Graphs showing the Piedmont, Coastal Plain, and NBSD data for the drainage-area categories of 100 and 200 mi^2, as well as the secondary envelope curves for those categories, are shown in figures 29 and 30. Summary comments on these secondary envelope curves follow.

Secondary Envelope Curve for Drainage Area of 100 Square Miles

The South Carolina live-bed contraction-scour data, in conjunction with the left-boundary point, described previously, provide a sufficient means to define the secondary envelope curve (fig. 29). The Coastal Plain data point that exceeds the secondary envelope curve is associated with a channel bend that will tend to produce larger than normal scour depth. The two NBSD data points for this category fall within the secondary envelope curves, providing limited support for the validity of the curve.

Table 5. Range of selected site characteristics for field measurements of contraction scour used to develop the modified live-bed contraction-scour envelope curve.

[mi^2, square mile; ft/ft, foot per foot; ft/s, foot per second; ft, feoo; mm, millimeter]

Range value	Drainage area (mi^2)	Channel slope (ft/ft)	Average approach channel velocity (ft/s)	Average approach channel depth (ft)	Average channel width (ft)	Geometric-contraction ratio	Median grain size (mm)	Measured contraction-scour depth (ft)
South Carolina Piedmont (Benedict and Caldwell, 2009) (24 measurements)								
Minimum	21	0.0005	2.4	7.7	41	0.14	0.51	0.0
Median	105	0.0012	5.5	13.7	72.5	0.64	0.77	3.7
Maximum	199	0.0021	8.8	18.1	210	0.92	1.7	9.4
South Carolina Coastal Plain (Benedict and Caldwell, 2009) (15 measurements)								
Minimum	17.2	0.00013	1.3	5.2	21	0.44	0.23	3.1
Median	124	0.00074	2.5	9.5	53	0.84	0.59	5.2
Maximum	198	0.002	7.1	15.8	140	0.93	1.7	17.1
South Carolina Clear-Water Contraction Scour (Benedict, 2003) (38 measurements)								
Minimum	6.1	0.0003	0.2[a]	3.1[b]	floodplain	0.77	0.06	3.2
Median	30.9	0.00076	0.5[a]	5.4[b]	floodplain	0.91	0.22	6.8
Maximum	120	0.0024	0.9[a]	8.0[b]	floodplain	0.97	0.78	12.8
National Bridge Scour Database (U.S. Geological Survey, 2001) (2 measurements)								
Minimum	10.3	0.001	0.7	13.8	42	0.95	0.1	4.0
Maximum	10.3	0.001	1.0	14.2	42	0.95	0.1	8.0

[a]These sites are associated with swampy floodplains with shallow, poorly defined channels; the average approach flow velocity was based on the average approach floodplain velocity.

[b]These sites are associated with swampy floodplains with shallow, poorly defined channels; the average approach flow depth was based on the average approach floodplain depth.

A total of 37 South Carolina clear-water contraction-scour data measurements were used for this category, and all but one fall within the secondary envelope curve, providing additional support for the validity of the curve.

Secondary Envelope Curve for Drainage Area of 200 Square Miles

The South Carolina live-bed contraction-scour data, in conjunction with the left-boundary point described previously, provide a sufficient means to define the secondary envelope curve (fig. 30). The Coastal Plain data point that slightly exceeds the secondary envelope curve is associated with a channel bend that will tend to produce larger than normal scour depth. Only one South Carolina clear-water contraction-scour measurement falls within this category providing limited support for the validity of the secondary envelope curve; no NBSD live-bed contraction-scour data measurements were available for this category.

Family of Curves for the Piedmont and Coastal Plain

The family of curves for live-bed contraction-scour depth in the Piedmont and Coastal Plain are shown in figure 31. Equations for the family of curves are shown in table 6.

Guidance and Limitations for Applying the Modified Contraction-Scour Envelope Curves

Benedict and Caldwell (2009) noted that substantial uncertainty can be associated with assessing the potential for live-bed contraction scour in South Carolina, making judgment an integral part in evaluating this component of scour. Although the modified South Carolina live-bed contraction-scour envelope curves (fig. 31; table 6) can be a useful supplementary tool for assessing live-bed contraction-scour at bridges in South Carolina, the difficulty in evaluating this scour component should be considered carefully. When applying

Table 6.   Equations for the modified live-bed contraction-scour envelope curve in the Piedmont and Coastal Plain of South Carolina.

[mi^2, square miles; <, less than; ≤, less than or equal to; DA, drainage area; y_S, scour depth, in feet; m, geometric-contraction ratio]

Drainage-area category	Equation	Limits of the geometric-contraction ratio
0 mi^2 < DA ≤ 100 mi^2	$y_S = 16\,m^2$	$0 \leq m \leq 0.9$
100 mi^2 < DA ≤ 200 mi^2	$y_S = 20\,m^2 + m$	$0 \leq m \leq 0.9$

the modified live-bed contraction-scour envelope curves, one must select a reference surface, estimate the drainage area and geometric-contraction ratio, select the appropriate contraction-scour envelope curve, evaluate other scour components in the contraction-scour region, and consider the limitations associated with the envelope curves. The following guidance may provide some assistance in applying the modified envelope curves.

1.   Select a reference surface:

The reference surface should be the average thalweg elevation along the profile of the channel in the live-bed contraction-scour region at the bridge. The thalweg is defined as the low point of the channel bed and should represent the natural conditions unaffected by scour. This reference surface can be determined by plotting the thalweg elevation at selected cross sections along the channel profile and then placing a best-fit line through those data to determine a reference surface. In many cases, defining the average thalweg elevation should not be a difficult task; however, the channel-bed topography at selected sites can be complex, making the determination of a reference surface more difficult. In such cases, judgment should be applied, bearing in mind that lower reference-surface elevations will produce lower scour-hole elevations and more conservative scour assessments. For additional details refer to Benedict and Caldwell (2009).

2.   Estimate of the drainage area and the geometric-contraction ratio:

The modified envelope curves can be sensitive to the selection of drainage area and the geometric-contraction ratio. Therefore, it is important that good estimates of these variables be obtained. The drainage area for a given site can be determined from use of standard computer software for determining topographic features. The geometric-contraction ratio can be determined by using flow models, topographic maps, and road plans, and when possible, all three sources should be used for verification. When discrepancies exist between these sources, judgment should be used to determine the most reasonable estimate of the geometric-contraction ratio.

3.   Select the appropriate live-bed contraction-scour envelope curve:

Live-bed contraction-scour depth can be assessed using the original envelope curve (Benedict and Caldwell 2009; fig. 24) or the modified envelope curves (fig. 31; table 6). The original and modified envelope curves do not make a distinction between the Piedmont and Coastal Plain; therefore, the primary criteria for selecting the appropriate envelope curve will be site characteristics for the bridge of interest. To utilize the modified envelope curves, the drainage area must be 200 square miles or less and the geometric-contraction ratio should not exceed 0.9. Additionally, the site characteristics for the bridge of interest should fall within the range of the appropriate South Carolina regional data used to develop the modified envelop curves as shown in table 5. If these criteria are not met, then the original live-bed contraction-scour envelope curve must be used (fig. 24) following the application guidance from Benedict and Caldwell (2009). Because of the uncertainty of assessing live-bed contraction scour, interpolation between the secondary envelope-curves for live-bed contraction scour is not recommended.

4.   Pier scour within the contraction-scour region:

Because of the complexity associated with pier and contraction scour in a river channel, Benedict and Caldwell (2009) concluded that it was uncertain if the original live-bed contraction-scour envelope curves (fig. 24) would account for pier scour. Therefore, judgment must be used to account for potential pier scour, in addition to the component of live-bed contraction scour when using the original or modified live-bed contraction scour envelope curves. For additional guidance refer to Benedict and Caldwell (2009).

5.   Limitations of the modified live-bed contraction-scour envelope curves:

The modified South Carolina live-bed contraction-scour envelope curves (fig. 31; table 6) can be useful supplementary tools for assessing live-bed contraction-scour potential at bridges in South Carolina. However, the following limitations of these empirical envelope curves should be considered.

- The modified envelope curves were developed from a limited sample of bridges in the Piedmont and Coastal Plain of South Carolina, and it is possible that scour depths could exceed the envelope curves. Therefore, applying a safety factor to the modified envelope curves may be prudent.

- Application of the modified envelope curves should be limited to bridges having site characteristics similar to the South Carolina data used to develop the envelope curves (table 5).

- Although clear-water contraction-scour data were used to verify the general trends of the live-bed contraction-scour data, the modified live-bed contraction-scour envelope curves should not be used to assess the potential for clear-water contraction scour.

- The modified envelope curves were developed using field data from sites assumed to have experienced flows near the 1-percent AEP (100-year) flow. Therefore, these curves should not be used to evaluate live-bed contraction-scour depths for larger flows, such as the 0.2-percent AEP (500-year) flow.

- The modified envelope curves do not account for adverse field conditions, such as channel bends and debris that may increase scour depths. For additional information refer to Benedict and Caldwell (2009).

Although the modified South Carolina live-bed contraction-scour envelope curves presented in this report can serve as a valuable supplementary tool in assessing live-bed contraction-scour depths in South Carolina, the noted limitations restrict their use. Therefore, the envelope curves should not be relied upon as the only tool for assessing live-bed contraction scour. To best assess anticipated scour, one should compile and study the available information for a given site and then bring sound engineering principles to bear on the final estimate of live-bed contraction-scour depth.

Summary

The U.S. Geological Survey, in cooperation with the South Carolina Department of Transportation, investigated historic scour at 231 bridges in the Piedmont and Coastal Plain physiographic provinces of South Carolina. These data were used to develop a suite of field-derived envelope curves that provided supplementary tools to assess scour potential in South Carolina for selected scour components, including clear-water abutment, contraction, and pier scour, as well as live-bed pier and contraction scour. The envelope curves consisted of a single curve with one explanatory variable that encompassed the measured field data for the respective scour components. In this investigation, the clear-water abutment-scour and live-bed contraction-scour envelope curves were modified to include a family of curves, providing a means to refine the assessment of scour potential for those scour components.

To determine an appropriate means to modify the original clear-water abutment-scour and live-bed contraction-scour envelope curves with a family of curves, patterns in selected laboratory and field data were reviewed. Based on these reviews, conceptual models for modifying the original envelope curves were developed. In the case of the clear-water abutment-scour envelope curves, it was concluded that for a constant embankment length, abutment-scour depth will increase at an increasing rate as the geometric-contraction ratio increases. Therefore, to develop a family of curves, the

South Carolina abutment-scour field data were grouped into categories of embankment length blocking flow and plotted against the geometric-contraction ratio. The application of this conceptual model led to the development of a series of five secondary envelope curves having embankment-length categories of 100-foot (ft) increments ranging from 100 to 500 ft. In the case of the live-bed contraction-scour envelope curve, it was concluded that the upper bound of contraction-scour depth will increase as the geometric-contraction ratio increases. Additionally, it was concluded that contraction scour is influenced by drainage area such that the upper bound of contraction-scour depth will increase as the drainage area increases. Therefore, to develop a family of curves, the South Carolina contraction-scour field data were grouped into categories of drainage area, and these were plotted against the geometric-contraction ratio. The application of this conceptual model led to the development of a series of two secondary envelope curves having drainage-area categories of 100 square-mile (mi^2) increments ranging from 100 to 200 mi^2.

The modified South Carolina clear-water abutment-scour and live-bed contraction-scour envelope curves can be useful supplementary tools for assessing scour potential at bridges in South Carolina. However, the limitations of the envelope curves must be carefully considered, and they should not be used at sites outside the range of data from which they were developed. Additional limitations and guidance for applying the modified envelope curves are provided in the report.

Selected References

Ballio, Francesco, Teruzzi, A., and Radice, A., 2009, Constriction effects in clear-water scour at abutments, Journal of Hydraulic Engineering, v. 135, no. 2, p. 140–145.

Benedict, S.T., 2003, Clear-water abutment and contraction scour in the Coastal Plain and Piedmont Provinces of South Carolina, 1996–99: U.S. Geological Survey Water-Resources Investigations Report 03–4064, 137 p.

Benedict, S.T., 2007, Development of regional envelope curves for assessing limits and trends in scour, *in* The World Environmental and Water Resources Congress 2007, Tampa, Florida, 2007, Proceedings: Reston, Virginia, American Society of Civil Engineers, p. 1–11.

Benedict, S.T., and Caldwell, A.W., 2006, Development and evaluation of clear-water pier and contraction scour envelope curves in the Coastal Plain and Piedmont Provinces of South Carolina: U.S. Geological Survey Scientific Investigations Report 2005–5289, 98 p.

Benedict, S.T., and Caldwell, A.W., 2009, Development and evaluation of live-bed pier and contraction scour envelope curves in the Coastal Plain and Piedmont Provinces of South Carolina, U.S. Geological Survey Scientific Investigations Report 2009–5099, 108 p.

Breusers, H.N.C., Nicollet, G., and Shen, H.W., 1977, Local scour around cylindrical piers: Journal of Hydraulic Research, v. 15, no. 3, p. 211–252.

Breusers, H.N.C., and Raudkivi, A.J., 1991, Scouring: Rotterdam, A.A. Balkema, 143 p.

Das, B.P., 1973, Bed scour at end-dump channel constrictions: Journal of the Hydraulics Division, v. 99, no. 12, p. 2273–2291.

Dey, S., and Raikar, R.V., 2005, Scour at long contractions, Journal of Hydraulic Engineering, v. 131, no. 12, p. 1036–1049.

Dongol, D.M.S., 1993, Local scour at bridge abutments: New Zealand, University of Auckland, School of Engineering Report no. 544, 410 p.

Ettema, R., Constantinescu, G., and Melville, B., 2011, Evaluation of bridge scour research: Pier scour processes and predictions: Transportation Research Board, National Cooperative Highway Research Program Web Document 175 (Project 24–27(01)), 195 p., accessed July 21, 2011, at *http://onlinepubs.trb.org/onlinepubs/nchrp/nchrp_w175.pdf*.

Ettema, R., Melville, B.W., and Barkdoll, Brian, 1998, Scale effect in pier-scour experiments: Journal of Hydraulic Engineering, v. 124, no. 6, p. 639–642.

Ettema, R., Yoon, B., Nakato, T., and Muste, M., 2004, A review of scour conditions and scour-estimation difficulties for bridge abutments: Journal of Civil Engineering, v. 8, no. 6, p. 643–650.

Ettema, R., Yorozuya, A., Nakato, T., and Muste, M., 2005, A practical approach to estimating realistic depths of abutment scour, *in* Proceedings of the 2005 Mid-Continent Transportation Research Symposium, Ames, Iowa, August 2005: Iowa State University.

Guimaraes, W.B., and Bohman, L.R., 1992, Techniques for estimating magnitude and frequency of floods in South Carolina, 1988: U.S. Geological Survey Water-Resources Investigations Report 91–4157, 174 p.

Hayes, D.C., 1996, Scour at bridges in Delaware, Maryland, and Virginia: U.S. Geological Survey Water-Resources Investigations Report 96–4089, 35 p.

Hurley, N.M., Jr., 1996, Assessment of scour-critical data collected at selected bridges and culverts in South Carolina, 1990–92: U.S. Geological Survey Open-File Report 96–350, 119 p.

Laursen, E.M., 1960, Scour at bridge crossings: Journal of the Hydraulics Division, American Society of Civil Engineers, v. 86, no. 2, p. 39–54.

Lombard, P.J., and Hodgkins, G.A., 2008, Comparison of observed and predicted abutment scour at selected bridges in Maine: U.S. Geological Survey Scientific Investigations Report 2008–5099, 23 p.

Melville, B.W., 1992, Local scour at bridge abutments: Journal of Hydraulic Engineering, American Society of Civil Engineering, v. 118, no. 4, p. 615–630.

Melville, B.W., and Coleman, S.E., 2000, Bridge scour: Highlands, Colorado, Water Resources Publications, 550 p.

Mueller, D.S., and Wagner, C.R., 2005, Field observations and evaluations of streambed scour at bridges: Federal Highway Administration, Publication FHWA-RD-03-052, 122 p.

Richardson, E.V., and Davis, S.R., 2001, Evaluating scour at bridges (4th ed.): Federal Highway Administration Hydraulic Engineering Circular No. 18, Publication No. FHWA NHI 01-001, 378 p.

Shearman, J.O., 1990, User's manual for WSPRO— A computer model for water-surface profile computations: Federal Highway Administration, Report No. FHWA-IP-89-027, 175 p.

U.S. Geological Survey, 2001, National bridge scour database: Accessed July 22, 2011, at *http://water.usgs.gov/osw/techniques/bs/BSDMS/index.htm*.

Wagner, C.R., Mueller, D.S., Parola, A.C., Hagerty, D.J., and Benedict, S.T., 2006, Scour at contracted bridges: Transportation Research Board, National Cooperative Highway Research Program Document 83 (Project 24–14), 299 p., accessed December 19, 2008, at *http://onlinepubs.trb.org/onlinepubs/nchrp/nchrp_w83.pdf*.

Zalants, M.G., 1990, Low-flow characteristics of natural streams in the Blue Ridge, Piedmont, and upper Coastal Plain Physiographic Provinces of South Carolina: U.S. Geological Survey Water-Resources Investigations Report 90–4188, 92 p.

Zalants, M.G., 1991, Low-flow frequency and flow duration of selected South Carolina streams through 1987: U.S. Geological Survey Water-Resources Investigations Report 91–4170, 87 p.

www.ingramcontent.com/pod-product-compliance
Lightning Source LLC
Chambersburg PA
CBHW081802170526

45167CB00008B/3286